Advances in Multifunctional Materials and Systems II

Advances in Multifunctional Materials and Systems II

Ceramic Transactions, Volume 245

A Collection of Papers Presented at the
10th Pacific Rim Conference on
Ceramic and Glass Technology
June 2–6, 2013
Coronado, California

Edited by
Jun Akedo
Tseung-Yuen Tseng
Xiang Ming Chen

Volume Editor
Hua-Tay Lin

The American Ceramic Society

WILEY

Published by John Wiley & Sons, Inc., Hoboken, New Jersey.
Published simultaneously in Canada.

For general information on our other products and services or for technical support, please contact our
Customer Care Department within the United States at (800) 762-2974, outside the United States at
(317) 572-3993 or fax (317) 572-4002.

Wiley also publishes its books in a variety of electronic formats. Some content that appears in print may
not be available in electronic formats. For more information about Wiley products, visit our web site at
www.wiley.com.

Library of Congress Cataloging-in-Publication Data is available.

ISBN: 978-1-118-77127-3
ISSN: 1042-1122

Printed in the United States of America.

10 9 8 7 6 5 4 3 2 1

Contents

OXIDE MATERIALS FOR NONVOLATILE MEMORY TECHNOLOGY AND APPLICATIONS

Preface

This Ceramic Transactions volume represents selected papers based on presentations in four symposia during the 10th Pacific Rim Conference on Ceramic and Glass Technology, June 2–6, 2013 in Coronado, California. The symposia include:

- Symposium 12: Advances in Electroceramics
- Symposium 13: Microwave Materials and Their Applications
- Sympoisum 14: Oxide Materials for Nonvolatile Memory Technology and Applications
- 2nd International Richard M. Fulrath Symposium on Frontiers of Ceramics for Sustainable Development

The editors wish to extend their gratitude and appreciation to all the co-organizers for their help and support, to all the authors for their cooperation and contributions, to all the participants and session chairs for their time and efforts, and to all the reviewers for their valuable comments and suggestions. Thanks are due to the staff of the meetings and publication departments of The American Ceramic Society for their invaluable assistance. We also acknowledge the skillful organization and leadership of Dr. Hua-Tay Lin, PACRIM 10 Program Chair.

JUN AKEDO, National Institute of Advanced Industrial Science and Technology, Japan
TSEUNG-YUEN TSENG, National Chiao Tung University, Taiwan
XIANG MING CHEN, Zhejiang University, China

Advances in
Electroceramics

PYROELECTRIC PERFORMANCES OF RELAXOR-BASED FERROELECTRIC SINGLE CRYSTALS AND THEIR APPLICATIONS IN INFRARED DETECTORS

Long Li,[1, 2] Haosu Luo,[1, *] Xiangyong Zhao,[1] Xiaobing Li,[1] Bo Ren,[1] Qing Xu,[1,2] and Wenning Di[1]

[1]Key Laboratory of Inorganic Functional Material and Device, Shanghai Institute of Ceramics, Chinese Academy of Sciences, 215 Chengbei Road, Jiading, Shanghai 201800, China
[2]University of Chinese Academy of Sciences, Beijing 100049, China

ABSTRACT

In this work the pyroelectric performances of relaxor-based ferroelectric single crystals (PMN-PT, Mn-doped PMN-PT, ternary PIN-PMN-PT and Mn-doped PIN-PMN-PT) are reported. The crystals show high pyroelectric coefficients (p), especially for PMN-0.26PT and Mn-doped PMN-0.26PT with values as high as 15.3×10^{-4} C/m^{2}K and 17.2×10^{-4} C/m^{2}K respectively. A co-design methodology of the macroscopic symmetry constraint controlling spontaneous polarization order parameter and dipole defects pinning controlling dynamic loss was established to control growth of crystals and reveal the physical mechanism of low dielectric loss for Mn-doped crystals. Dielectric losses of binary and ternary relaxor-based single crystals are depressed to 0.05 %, enhancing the detectivity figure of merit (F_d) up to 40.2×10^{-5} Pa$^{-1/2}$ for Mn-doped PMN-0.26PT. The simulations were carried out for the performances of relaxor-based single crystal detectors in order to fabiricate high performance detectors. The results show that relaxor-based ferroelectric single crystals have great advantages compared with the conventional LiTaO$_3$ and DTGS in the low frequency range. By using Mn-doped PMNT single crystals as sensitive element and the multi-walled carbon nanotubes as absorbing layer, the outstanding infrared detectors were achieved. The specific detectivity (D^*) of Mn-doped PMNT-based detector is up to 3.01×10^9 cmHz$^{1/2}$/W (at 2 Hz) and 2.21×10^9 cmHz$^{1/2}$/W (10 Hz, 500 K, 25 °C) respectively, four times higher than that of LiTaO$_3$-based detectors.

INTRODUCTION

Pyroelectric infrared detectors exhibit advantages for the wide wavelength response, uncooled, high sensitivity, compacted structures and low cost, which are enable a variety of applications such as body detectors, flame and fire detectors, IR spectrometry, gas analyzers, night vision, thermal imaging and IR camera.[1,2] Nowadays traditional pyroelectric bulk materials, such as triglycine sulfate (TGS), lithium tantalate (LiTaO$_3$), barium strontium titanate (BST), and lead scandium tantalate (PST) have been widely utilized for fabricating single-element infrared (IR) sensors and portable uncooled IR focal plane arrays (UFPAs) in military, paramilitary and commercial imaging applications.[3,4] To enhance the performances of the IR detectors, much attention has been paid on the novel pyroelectric materials with high pyroelectric coefficient, high figure of merits (FOMs), low dielectric loss and high temperature stability.[1,5]

In 1996, large-size single crystals of PMNT were fabricated and then reported.[6,7] In 2002, Zhao discovered their high pyroelectric coefficients along the spontaneous polarization direction of [111] in rhombohedral phase.[8] From then on, related researches on structure, composition, orientation and performances of the crystals have been carried out in detail.[9-17] All these results show that relaxor-based ferroelectric single crystals are promising for the next-generation high performance pyroelectric materials used for IR detection applications.

In this paper, we present our recent work and progresses on the fabrication and properties in these novel pyroelectric materials of pure PMNT, PIMNT, Mn-doped PMNT and Mn-doped PIMNT. More important is that we report our main progress in the fabrication of high performance IR detectors using Mn-doped PMNT single crystals.

EXPERIMENTS AND RESULTS

(1) Growth and Properties of Single Crystals

Relaxor-based ferroelectric single crystals PMNT, Mn-doped PMNT, PIMNT and Mn-doped PIMNT (Figure 1) were all grown by the modified Bridgman technique.[7] The crystals were oriented along [001], [110] and [111] directions using an x-ray diffractometer.[18] For the characterization, the crystals were cut into specimens with dimensions of 4×4×0.5 mm^3,[19] coated with silver paste and sintered at 700 °C. Then, the samples were poled under an electric field of 4×E_c for 15 min at a high temperature in silicone oil. The pyroelectric coefficient was measured by a dynamic technique using sinusoidal temperature change at very low frequency of 45 mHz. The dielectric properties were performed with a HP4294 impedance analyzer while the hysteresis loops were measured with a TF1000.

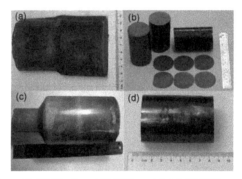

Figure 1. (a) PMNT crystal, (b) Mn-doped PMNT crystal, (c) PIMNT crystal and (d) Mn-doped PIMNT crystal.

By analysis of relation between pyroelectric properties and compositions, phase structures and crystallographic directions optimization of compositions and structures, growth technology of high performance and large-size relaxor-based ferroelectric single crystals was established and optimized. The PT-content (x) dependence of pyroelectric coefficients (p) for the crystals at room temperature is shown in Figure 2 (a).[15] The pyroelectric coefficients increase intensively with the decrease of x and exhibit their largest values along spontaneous polarization directions where the structures of the crystals are rhombohedral ([111]), morphotropic phase boundary ([110]) and tetragonal ([001]) for $x\leq0.30$, $0.30<x<0.35$ and $x\geq0.35$, respectively.[20] Temperature dependence of the pyroelectric coefficient for the crystals is given in Figure 2 (b). The values of

p for PMN-0.26PT, Mn-PMN-0.26PT, PIMNT (41/17/42) and Mn-PIMNT (23/47/30) crystals at room temperature are 15.3×10^{-4} C/m²K, 17.2×10^{-4} C/m²K, 5.7×10^{-4} C/m²K and 7.37×10^{-4} C/m²K, respectively, much higher than those of traditional pyroelectric materials (LiTaO₃ and PZT). As the temperature increases, the pyroelectric coefficients increase slightly.

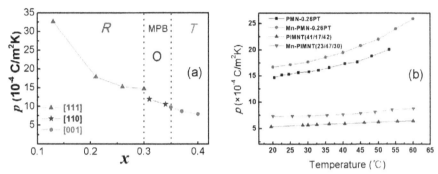

Figure 2. (a) Composition dependence of the pyroelectric coefficient for PMN-xPT crystals at room temperature.[15] (b) Temperature dependence of pyroelectric coefficients for pure PMN-0.26PT,[25] Mn-doped PMN-0.26PT,[11] PIMNT (41/17/42)[12] and Mn-doped PIMNT (23/47/30) crystals along their spontaneous polarization directions.

To select pyroelectric materials for practical devices, three major figures of merits (FOMs) are introduced: current responsivity (F_i), voltage responsivity (F_v) and detectivity (F_d),[1] where the detectivity FOM (F_d) can be defined as:

$$F_d = p / [C_v (\varepsilon_0 \varepsilon_r \tan \delta)^{1/2}] \qquad (1)$$

Here, $p / \varepsilon_r^{1/2}$ is associated with the intrinsic parameters of the pyroelectric materials. For the crystals having the same structure and orientation, $p / \varepsilon_r^{1/2}$ has little change,[21-24] shown in Figure 3 (a), demonstrating that the detectivity FOM (F_d) can't be enhanced by improving the $p / \varepsilon_r^{1/2}$. So a co-design methodology was established, consisting of the macroscopic symmetry constraint controlling spontaneous polarization order parameter and dipole defects pinning controlling dynamic loss. In this work, Mn ions, occupying the B-site of perovskite structure, were introduced. The generated $\left(Mn_{Ti}^{2+} \right)'' - V_o^{\bullet\bullet}$ dipole defects can pinning the domain walls and suppress the transport of vacancy conductance,[26] revealing the physical mechanism of low dielectric loss for Mn-doped PMNT single crystals. The frequency dependence of dielectric properties in the range of 50 Hz to 10 kHz for pure and Mn-doped crystals is shown in Figure 3 (b). Very low dielectric losses of 0.05% and 0.049% at 1 kHz were observed for Mn-doped PMN-0.26PT and Mn-doped PIMNT (23/47/30), respectively. It is obvious that Mn-doping is an effective solution to decrease the dielectric losses for both PMNT and PIMNT crystals.

To extend the processing and operating temperature range, ternary PIN-PMN-PT single crystals were grown, announcing not only high Curie temperature but also good pyroelectric properties. The temperature dependence of dielectric constants is measured in Figure 4 for pure PMN-0.26PT, PIMNT (41/17/42), Mn-doped PMN-0.26PT and Mn-doped PIMNT (23/47/30) single crystals. The Curie temperatures of PIMNT (41/17/42) and Mn-doped PIMNT (23/47/30) are shifted to 253 °C and 179 °C, respectively, outperforming that of PMN-0.26PT crystals (125 °C) significantly.

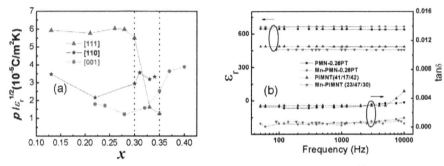

Figure 3. (a) $p/\varepsilon_r^{1/2}$ of PMN-xPT single crystals as a function of composition,[25] (b) Frequency dependence of dielectric properties for pure PMN-0.26PT, Mn-doped PMN-0.26PT, PIMNT (41/17/42)[11,12,25] and Mn-doped PIMNT (23/47/30) crystals poled along their spontaneous polarization directions.

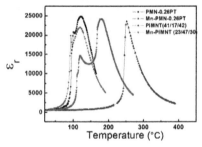

Figure 4. Temperature dependence of dielectric constants for pure PMN-0.26PT, Mn-doped PMN-0.26PT, PIMNT (41/17/42)[11,12,25] and Mn-doped PIMNT (23/47/30) crystals.

Hysteresis loops for PMN-0.26PT, Mn-doped PMN-0.26PT, PIMNT (41/17/42) and Mn-doped PIMNT (23/47/30) single crystals were measured at room temperature (shown in Figure 5). The remnant polarizations (P_r) of the crystals are 35 μC/cm², 42 μC/cm², 33 μC/cm² and 41 μC/cm², respectively. Here, Mn-doped PMN-0.26PT crystal has a coercive field (E_c) of 4.5 kV/cm, larger than that of pure PMN-0.26PT (2.6 kV/cm), and its P-E loop shows a positive offset of 0.5 kV contributed by oxygen vacancy defect-dipoles.[26] Furthermore, the E_c of 16 kV/cm and 8 kV/cm for PIMNT (41/17/42) and Mn-doped PIMNT (23/47/30) respectively have been enhanced large enough to keep the crystals stable when an inverse electric field is applied in the poling direction.

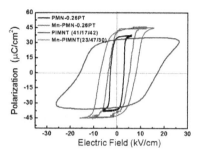

Figure 5. P-E loops of PMN-0.26PT, Mn-doped PMN-0.26PT, PIMNT (41/17/42)[11,12,25] and Mn-doped PIMNT (23/47/30) crystals at room temperature.

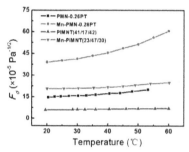

Figure 6. Temperature dependence of the detectivity FOM (F_d) for PMN-0.26PT, Mn-doped PMN-0.26PT, PIMNT (41/17/42) and Mn-doped PIMNT (23/47/30) crystals.

Table I. Pyroelectric and related properties of relaxor-based ferroelectric single crystals and other typical pyroelectric materials

Pyroelectric materials	T_C °C	p $10^{-4}C/m^2K$	ε_r	$\tan\delta$ @1kHz	F_i $10^{-10}m/V$	F_v m^2/C	F_d $10^{-5}Pa^{-1/2}$	Ref.
PMN-0.26PT	122	15.3	643	0.0028	6.1	0.11	15.3	25
Mn-PMN-0.26PT	120	17.2	660	0.0005	6.88	0.12	40.2	11
PMN-0.29PT	135	12.8	515	0.0063	5.25	0.11	9.8	10
Mn-PMN-0.29PT	135	16.2	688	0.0005	6.48	0.11	37.1	28
PIMNT (21/49/30)	180	7.54	529	0.001	2.8	0.06	13	17
PIMNT (41/17/42)	253	5.7	487	0.003	2.28	0.05	6.3	12
Mn-PIMNT (23/47/30)	179	7.37	461	0.00049	2.75	0.067	19.5	This work
LiTaO₃	620	2.3	47	0.0005	0.72	0.17	15.7	15
PZT	340	3.3	714	0.02	1.32	0.02	1.17	15

The F_d temperature dependence was calculated to evaluate the performances of four kinds of relaxor-based ferroelectric crystals (Figure 6). Table I summarizes the pyroelectric and related properties of relaxor-based ferroelectric single crystals and other typical pyroelectric materials. At room temperature, the values of F_d are up to 40.2×10^{-5} Pa$^{-1/2}$ and 19.5×10^{-5} Pa$^{-1/2}$ for Mn-doped PMN-0.26PT and Mn-doped PIMNT (23/47/30), respectively, much higher than those of LiTaO$_3$ and PZT.[27]

(2) Physical Model and Performance Simulation of Devices

To provide instructions for detector preparation and improvement, the theoretical model and simulations were carried out. Thermal model of the sensitive element for the pyroelectric detector is given in Figure 7. When temperature of the chip is changed, there will be a release of charges at the surface of pyroelectric materials. The temperature change can be generated by electromagnetic radiation defined as:

$$P(t) = P_0 \exp(j\omega t) \qquad (2)$$

The charges can be detected as a current, i_p, flowing in an external circuit such that:

$$i_p = A_s p \, dT / dt \qquad (3)$$

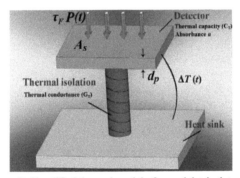

Figure 7. Simplified thermal model of pyroelelctric detector.

Taking both the thermal and current mode electrical circuit into consideration, we can describe the voltage responsivity (R_v) and specific detectivity (D^*) by the following equations:

$$R_v = \frac{A_s p \omega \alpha}{G\sqrt{1 + \omega^2 \tau_T^2}} \times \frac{R_f}{\sqrt{1 + \omega^2 \tau_E^2}} \qquad (4)$$

$$D^* = \frac{R_v (A_s \Delta f)^{1/2}}{V_n} \qquad (5)$$

The noise sources for the pyroelectric detectors are temperature or radiation noise, Johnson noise (consisting of resistor noise and dielectric loss noise), the preamplifier current noise and the preamplifier voltage noise.[1]

According to the above equations (2-5), the frequency dependence of D^* for each part of the noises was calculated and simulated (Figure 8). At low frequencies (<10 Hz), the current noise of the preamplifier dominates while at high frequencies (>100 Hz) the voltage noise of the preamplifier specifies the resultant noise density. In the range of 10 Hz to 100 Hz, the tanδ noise of sensitive element and the voltage noise of the preamplifier dominate together. So Mn-doped PMNT single crystals can be excellent materials to decrease the tanδ noise and improve the specific detectivity due to their low dielectric loss.

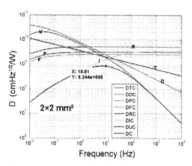

Figure 8. Detectivity of various components for the device noises and frequency dependence of the resultant D^*.

The frequency dependence of D^* for PMNT-based detectors with different responsive area is shown in Figure 9. For single-element detectors ($A_s > 1 \times 1$ mm^2), the maximum values of D^* are obtained between 1 Hz and 10 Hz, and for focal plane array detectors ($A_s < 100 \times 100$ μm^2), the maximum values of D^* are achieved between 100 Hz and 1 kHz, which provides a basis for responsive area choosing in different application objects. Furthermore, the chip thickness dependence of D^* for PMNT-based detectors was simulated. For single-element detectors (Figure 10 (a)), the D^* increases significantly as the chip thickness is reduced from 30 μm to 5 μm at the low frequencies (<10 Hz), and at the relatively high frequencies (>10 Hz), the D^* remains basically unchange. For focal plane array detectors, the D^* can be improved in a very wide frequency range by chip thickness reduction, see Figure 10 (b).

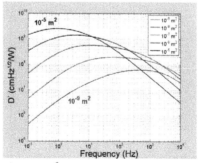

Figure 9. Frequency dependence of D^* for PMNT-based detectors with different sensitive areas.

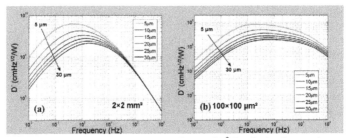

Figure 10. Thickness of sensitive elements dependence of D^* for single-element detectors (a) and focal plane array detectors (b).

The comparison of D^* calculated from the theoretical responsivity (R_v) and noise (V_n) for detectors based on different pyroelectric materials (current mode) is presented in Figure 11. Due to the higher responsivity relying on high pyroelectric coefficient and the lower noise relying on low dielectric loss, PMNT-based detectors are distinguished by several times higher detectivity over LiTaO$_3$-based and DTGS-based detectors in the low frequency range (f<100 Hz). Higher detectivity of about 4×10^9 cmHz$^{1/2}$/W (at 3 Hz) and 3.5×10^9 cmHz$^{1/2}$/W (at 10 Hz) can be obtained for detectors based on Mn-doped PMN-0.26PT single crystals.

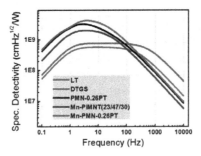

Figure 11. Frequency dependence of theoretical specific detectivity (D^*) for different pyroelectric materials-based detectors.

(3) Detector Design and Measurement

The performances of pyroelectric detectors are closely related not only to the properties of the pyroelectric materials, but also to the size and thickness of sensitive elements, the absorbing layers, the matching circuitry and the resultant noises. Here we selected Mn-doped PMNT single crystals for IR detector fabrication because of their high pyroelectric coefficient and low dielectric loss. For sensitive elements, thinning and polishing technique of large-size crystal wafers was established, and different absorbing materials were also studied and compared. To exploit advantages of relaxor-based ferroelectric single crystals, a current mode circuitry with matching electronic components was established. Solution of all these critical problems can significantly improve performances of the pyroelectric detectors.

The basic structure of pyroelectric detector is depicted in Figure 12. The single crystals were thininged and polished to 20 μm in thickness, and then the electrodes were deposited on top and bottom surfaces of the crystals by magnetron sputtering. The sensitive area of the chip is the cross section between Ni-Cr electrode (50 nm) and Ni-Cr/Au electrode (200 nm). To improve the absorption of infrared radiation, an absorbing layer must be sprayed on the top electrode. Traditionally, Ni-Cr alloy, silver black and gold black are utilized as the coating materials, but they have the drawbacks of low absorptivity or high cost. In this work we selected a kind of multi-walled carbon nanotubes,[29] exhibiting high absorptivity (99%), large thermal conductivity (2000 W/(mK)) and low cost, as the coating materials. Finally the detector was mounted in a metal package with pyroelectric chip free standing. The fabricated Mn-doped PMNT-based infrared detectors with different windows for different applications are shown in Figure 13 (a)-(d). Figure 13 (e) presents the matching circuitry for performance measurement.

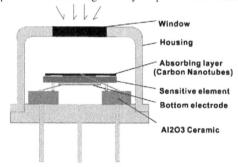

Figure 12. The basic structure of pyroelectric detector.

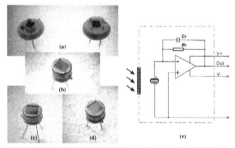

Figure 13. (a)-(d) As-prepared Mn-doped PMNT-based pyroelectric detectors, (e) Current mode matching circuitry for the detector.

Responsivity (R_v) of the detector in the frequency range of 0.5 Hz to 1 kHz was measured with a blackbody, running at 500 K (shown in Figure 14). The signal voltages of 0.9 Hz and 10 Hz reach the values of 3000 mV, 886 mV, respectively, promoting the R_v up to 80355 V/W at 10 Hz and showing an inverse proportion to the modulation frequency above the electrical corner frequency (f_E=2.6 Hz).

Outstanding specific detectivity (D^*) in the low frequency range is achieved for Mn-doped PMNT-based infrared detector. Modulation frequency dependence of specific detectivity calculated from the measurement signal and noise voltages is presented in Figure 15. At 10 Hz, the D^* is up to 2.21×10^9 cmHz$^{1/2}$/W with a noise of 6.45 μV/Hz$^{1/2}$. Table II summarizes the performances of different materials-based detectors. Mn-doped PMNT-based detector fabricated in this work performs much more excellent detectivity, 4 fold higher than that of commercial LiTaO$_3$-based detectors.

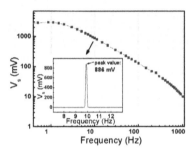

Figure 14. Modulation frequency dependence of responsivity (V_s) for the as-prepared detector (current mode).

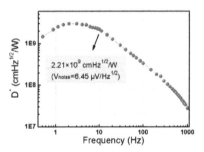

Figure 15. Modulation frequency dependence of specific detectivity (D^*) for Mn-doped PMNT-based detector (current mode).

Table II. Performances of different materials-based detectors

Parameter (10 Hz)	d μm	As mm^2	V_s mV	V_n μV/Hz$^{1/2}$	R_v V/W	D^* 10^8 cmHz$^{1/2}$/W	Ref.
Mn-PMNT	20	3.14	886	6.45	80355	22.1	This work
LiTaO$_3$	27	4	3.23	0.85	2308	5.4	30

CONCLUSION

In summary, pyroelectric performances of relaxor-based ferroelectric single crystals (PMN-PT, Mn-doped PMN-PT, ternary PIN-PMN-PT and Mn-doped PIN-PMN-PT) were evaluated. High pyroelectric coefficients were observed in the crystals along their spontaneous polarization directions, especially for rhombohedral phase PMNT with PT contents of 26%-30%. Low dielectric losses of single crystals were obtained by Mn ions doping, enhancing the detectivity figure of merit (F_d) up to 40.2×10^{-5} $Pa^{-1/2}$ for Mn-doped PMN-0.26PT. The enhancement of Curie temperature for ternary crystals, 253 $^\circ$C and 179 $^\circ$C for PIMNT (41/17/42) and Mn-doped PIMNT (23/47/30) respectively, enable much better temperature stability and larger operating temperature range. The results of simulations can provide reliable instructions for us to prepare and improve detectors. From the theoretical comparison of different materials-based detectors, the relaxor-based ferroelectric single crystals have great advantages over LTO and DTGS in the low frequency range and very suitable for fabrication of IR detectors. By using Mn-doped PMNT single crystals and multi-walled carbon nanotubes absorbing layer, an outstanding pyroelectric single-element detector was achieved. The specific detectivity D^* (10 Hz, 500 K, 25 $^\circ$C) of the detector was 2.21×10^9 $cmHz^{1/2}/W$, outperforming $LiTaO_3$-based detectors by 4 times more.

ACKNOWLEDGMENTS

This work was financially supported by the Ministry of Science and Technology of China through 973 Program (Nos. 2013CB632902-3, 2009CB623305, 2013CB6329052 and 2013CB632906), the Natural Science Foundation of China (Nos. 51332009, 51372258, 11304333, 61001041, 11090332 and 51272268), Science and Technology Commission of Shanghai Municipality (No. 12DZ0501000), Open Project from Shanghai Institute of Technical Physics, CAS (No. IIMDKFJJ-11-08), the Fund of Shanghai Institute of Ceramics (No. Y29ZC4140G and Y39ZC4140G), and Shanghai Municipal Electric Power Company (No. 52091413502W).

FOOTNOTES

[*]Author to whom correspondence should be addressed. e-mail: hsluo@mail.sic.ac.cn

REFERENCES

[1]R. W. Whatmore, "pyroelectric devices and materials," *Rep. Prog. Phys.*, **49** 1335–1386 (1986).

[2]P. W. Kruse, *Uncooled Thermal Imaging, Arrays, Systems, and Applications*, (SPIE press), **51** 49–55 (2001).

[3]R. Watton, "Ferroelectric IR bolometers – from ceramic hybrid arrays to direct thin film integration," *Ferroelectrics*, **184** 141–150 (1996).

[4]R. G. Buser, M. F. Tompsett, P. W. Kruse, R. A. Wood, C. M. Hanson, D. L. Polla, J. R. Choi, N. Teranishi, T. W. Kenny, J. R. Vig, R. L. Filler, and Y. Kim, *Uncooled Infrared Imaging Arrays and Systems*, (Columbus USA: ACADEMIC PRESS), **47** (1997).

[5]J. Brady, T. Schimert, and D. Ratcliff, "Advances in Amorphous Silicon Uncooled IR Systems," *Proc. SPIE* 3698, *Infrared Technology and Applications XXV*, **161** (1999).

[6]罗豪甦, 沈关顺, 王评初, 乐秀宏, 殷之文, "新型压电材料–弛豫铁电单晶的研究," *无机材*

料学报, **12** [5] (1997).

[7]H. Luo, G. Xu, H. Xu, P. Wang, and Z. Yin, "Compositional homogeneity and electrical properties of lead magnesium niobate titanate single crystals grown by a modified Bridgman technique," *Jpn. J. Appl. Phys.*, **39** [9B] 5581–5585 (2000).

[8]X. Zhao, B. Fang, H. Cao, Y. Guo, and H. Luo, "Dielectric and piezoelectric performance of PMN-PT single crystals with compositions around the MPB: influence of composition, poling field and crystal orientation," *Mater. Sci. Eng. B*, **96** [3] 254–262 (2002).

[9]X. Wan, H. Luo, J. Wang, H. L. W. Chan, and C. L. Choy, "Investigation on optical transmission spectra of $(1-x)Pb(Mg_{1/3}Nb_{2/3})O_3-xPbTiO_3$ single crystals," *Solid State Commun.*, **129** [6] 401–405 (2004).

[10]Y. Tang, X. Zhao, X. Feng, W. Jin, and H. Luo, "Pyroelectric properties of [111]-oriented $Pb(Mg_{1/3}Nb_{2/3})O_3$-$PbTiO_3$ crystals," *Appl. Phys. Lett.*, **86** [8] 082901 (2005).

[11]L. Liu, X. Li, X. Wu, Y. Wang, W. Di, D. Lin, X. Zhao, H. Luo, and N. Neumann, "Dielectric, ferroelectric, and pyroelectric characterization of Mn-doped $0.74Pb(Mg_{1/3}Nb_{2/3})O_3$-$0.26PbTiO_3$ crystals for infrared detection applications," *Appl. Phys. Lett.*, **95** [19] 192903 (2009).

[12]P. Yu, F. Wang, D. Zhou, W. Ge, X. Zhao, H. Luo, J. Sun, X. Meng, and J. Chu, "Growth and pyroelectric properties of high Curie temperature relaxor-based ferroelectric $Pb(In(1/2)Nb(1/2))O(3)$-$Pb(Mg(1/3)Nb(2/3))O(3)$-$PbTiO(3)$ ternary single crystal," *Appl. Phys. Lett.*, **92** [25] 252907 (2008).

[13]Y. Guo, H. Luo, T. He, and Z. Yin, "Peculiar properties of a high Curie temperature $Pb(In1/2Nb1/2)O3$-$PbTiO3$ single crystal grown by the modified Bridgman technique," *Solid State Commun.*, **123** [9] 417–420 (2002).

[14]Y. Tang, H. Luo, X. Zhao, H. Xu, T. He, D. Lin, and W. Jin, "Composition, dc bias and temperature dependence of pyroelectric properties of <111>-oriented $(1-x)Pb(Mg1/3Nb2/3)O3$-$xPbTiO3$ crystals," *Mater. Sci. Eng. B*, **119** 71–74 (2006).

[15]Y. Tang, X. Wan, X. Zhao, X. Pan, D. Lin, and H. Luo, "Large pyroelectric response in relaxor-based ferroelectric $(1-x)Pb(Mg1/3Nb2/3)O$-3-$xPbTiO(3)$ single crystals," *J. Appl. Phys.*, **98** [8] 084104 (2005).

[16]L. Liu, X. Wu, X. Zhao, X. Feng, W. Jing, and H. Luo, "Pyroelectric Performances of Rhombohedral $0.42Pb(In1/2Nb1/2)O3$-$0.3Pb(Mg1/3Nb2/3)O3$-$0.28PbTiO3$ Single Crystals," *IEEE Trans. Ultrason. Ferroelectr. Freq. Control.*, **57** [10] 2154–2158 (2010).

[17]L. Liu, X. Wu, S. Wang, W. Di, D. Lin, X. Zhao and H. Luo, "Growth and pyroelectric properties of rhombohedral $0.21Pb(In1/2Nb1/2)O3$-$0.49Pb(Mg1/3Nb2/3)O3$-$0.3PbTiO3$ ternary singlecrystals," *Journal of Crystal Growth*, **318** 856–859 (2011).

[18]M. Orita, H. Satoh, K. Aizawa, and K. Ametani, "Preparation of Ferroelectric Relaxor $Pb(Zn1/2Nb2/3)O3$–$Pb(Mg1/3Nb2/3)O3$–$PbTiO3$ by Two-Step Calcination Method," *Jpn. J. Appl. Phys.*, **31** 3261–4 (1992).

[19]M. Davis, D. Damjanovic, and N. Setter, "Pyroelectric properties of $(1-x)Pb(Mg_{1/3}Nb_{2/3})O_3-xPbTiO_3$ and $(1-x)Pb(Zn_{1/3}Nb_{2/3})O_3-xPbTiO_3$ single crystals measured using a dynamic method," *J. Appl. Phys.*, **96** [5] 2811 (2004).

[20]A. K. Singh and D. Pandey, "Structure and the location of the morphotropic phase boundary region in $(1-x)[Pb(Mg1/3Nb2/3)O3]$–$xPbTiO3$," *J. Phys.: Condens. Matter*, **13**, 931–936 (2001).

[21]S. T. Liu, J. D. Zook, and D. Long, "Relationships between pyroelectric and ferroelectrics parameters," *Ferroelectrics*, **9** 39–43 (1975).

[22]J. D. Zook and S. T. Liu, "Use of effective field theory to predict relationships among ferroelectric parameters," *Ferroelectrics*, **11** [1] 371-376 (1976).

[23]A. F. Devonshire, "Theory of ferroelectrics," *Advances in Physics*, **43** 85-130 (1954).

[24]M.E. Lines and A.M. Glass, "Principles and Applications of Ferroelectrics and Related Materials," *New York: Oxford University Press*, 2001.

[25]Y. Tang, "Novel pyroelectric materials and their applications in infrared devices," Ph. D thesis, Shanghai Institute of Ceramics, Chinese Academy of Sciences, 2007.

[26]G. E. Pike, W. L. Warren, D. Dimos, B. A. Tuttle, R. Ramesh, J. Lee, V. G.Keramidas, and J. T. Evans, "Voltage offsets in (Pb,La)(Zr,Ti)O₃ thin films," *Appl. Phys. Lett.*, **66** 484 (1995).

[27]B. M. Kulwicki, A. Amin, H. R. Beratan, and C. M. Hanson, "pyroelectric imaging," *Proc. 8th Int. Symp. on Applications of Ferroelectrics* (IEEE, New York), (1992).

[28]Y. Tang, L. Luo, Y. Jia, H. Luo, X. Zhao, H. Xu, D. Lin, J. Sun, X. Meng, J. Zhu, and M. Es-Souni, "Mn-doped 0.71Pb,,Mg1/3Nb2/3...O3–0.29PbTiO3 pyroelectric crystals for uncooled infrared focal plane arrays applications," *Appl. Phys. Lett.*, **89** 162906 (2006).

[29]X. Shao, X. Ma, Y. Yu, and J. Fang, "The study of carbon nanotubes as coating films for electrically calibrated detectors," *Meas. Sci. Technol.*, **23** 025106 (2012).

[30]N. Neumann, M. Es-Souni, and H. Luo, "Application of PMN-PT in pyroelectric detectors," *18th IEEE International Symposium on the Applications of Ferroelectrics*, (2009).

FORMATION OF TOUGH FOUNDATION LAYER FOR ELECTRICAL PLATING ON INSULATOR USING AEROSOL DEPOSITION METHOD OF Cu-Al$_2$O$_3$ MIXED POWDER

Naoki Seto, Shingo Hirose, Hiroki Tsuda and Jun Akedo

National Institute of Advanced Industrial Science &Technology

Tsukuba, Ibaraki, Japan

ABSTRACT

In electrical industry, it is important technology that making the electrode on insulator to make higher density electric circuits.   Especially in power electronics, it is quite important thick Cu plating for big electric current.   However adhesion of Cu plating is decreased because of increasing internal stress caused by increasing plating thickness.   The decrease of plating adhesion means reliability fall of power electric circuits.   Therefore tough and thick Cu plating on insulator has big interest from power electric industry.   However it is hard to achieve making tough and thick plating on insulator by easy and low cost method.

On the other hands, Aerosol Deposition Method has attractive features such as simple mechanism or combination flexibility between powder and substrate.   And electrical plating is famous method to make the electrode.   However it is hard to make electrode on insulator for plating, because substrate for plating must need conductivity.   Therefore, if tough and high conductivity coating can form on insulator using AD process, we can make the electrode on insulator in quite low cost.

In this report, authors investigated about formation of tough seed layer for plating on insulator by Cu-Al$_2$O$_3$ mixed powder using AD process.   And authors success not only making tough seed layer but also electric plating on seed layer.

INTRODUCTION

To make tough layer with low resistance and high conductivity on insulator is quite important for power electronics.   And such a characteristic layer is also important as making technique of high efficiency heat dissipation substrate.   The authors try to perform electrical plating on insulator in this study, however, making tough and low resistance layer is first barrier to achieve electrical plating on insulator.   Therefore, this study will be useful and important not only plating on insulator but also power electronics etc.   Such a background therefore, the authors investigated by following 2 steps

   1. How to make tough and low resistance seed layer on insulator.
    (And checking characteristics of seed layer)
   2. Formation trial of electrical plating on seed layer.
    (And checking characteristics of plating layer)

To investigate electric plating on insulator, the authors selected glass as substrate because glass is famous insulator.   However, electrical plating cannot perform on glass directly because insulator prevents electric current flow.   Therefore to get toughness and high conductivity, the authors select Al$_2$O$_3$ and Cu as seed layer materials.

EXPERIMENTAL SETUP

Aerosol Deposition Method (ADM) can make hard film of Al$_2$O$_3$[1] or composite material film[2] easily.   This characteristic of ADM is quite convenient for this study.   Therefore, the authors decided to use ADM to make foundation coating.   Figure 1 shows schematic drawing of ADM in this study.

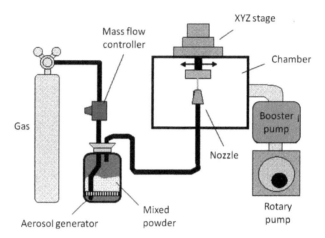

Figure 1 Schematic drawing of Aerosol Deposition Method

It is well known that Al_2O_3 as hard and tough material, and Cu as low electrical resistance material. Therefore the authors made $Cu-Al_2O_3$ mixed powder and put it in aerosol generator. In this experiment, the specific electrical resistance of seed layer is quite important because only low resistance can perform electrical plating on seed layer. Therefore we measured specific electrical resistance after making seed layer, and change the mixing rate of $Cu-Al_2O_3$ to optimize resistance.

After coating seed layer by AD process, the authors performed electrical Cu-plating on seed layer. Figure 2 shows electrical plating setup.

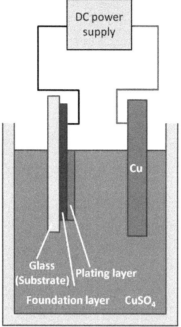

Figure 2 Schematic drawing of electrical plating setup

During plating, the authors observed change of seed layer carefully. And after performing the plating, the authors checked plating properties.

RESULTS AND DISCUSSIONS
At first, the authors tried to coat seed layer using AD process. Figure 3 shows typical result of seed layer.

Figure 3 Appearance of seed layer using AD process

The thickness of this seed layer is about 1.4μm, and the specific electrical resistance of this layers is about $3.1 \times 10^{-5} \Omega \cdot cm$. This resistance value is enough low to perform electrical plating. But before plating experiment, we investigated relation between specific electrical resistance and Cu-Al$_2$O$_3$ mixing rate to control electrical resistance of seed layer. Figure 4 shows relation between specific electrical resistance and Cu-Al$_2$O$_3$ mixing rate.

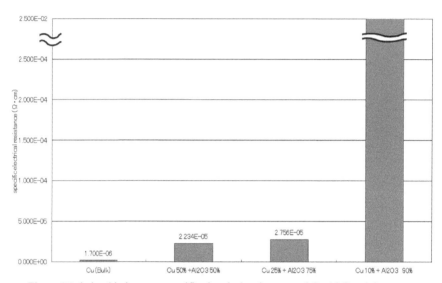

Figure 4 Relationship between specific electrical resistance and Cu-Al$_2$O$_3$ mixing rate

Bulk Cu was quite low resistance like $1.7 \times 10^{-6} \Omega \cdot$cm. And seed layer (Cu 50% + Al$_2$O$_3$ 50%) was higher resistance than bulk Cu, but this value is enough low to perform electrical plating. The case of Cu 25% + Al$_2$O$_3$ 75% seed layer, resistance is almost same to Cu 50% + Al$_2$O$_3$ 50% layer. However, the case of Cu 10% + Al$_2$O$_3$ 90%, resistance increased suddenly. Over $2.5 \times 10^{-2} \Omega \cdot$cm is hard to perform plating, therefore we decided that mixing rate of Cu 10% + Al$_2$O$_3$ 90% did not use for making seed layer.

And we also investigate toughness of seed layer by scratching directly. Figure 5 shows scratch test result of Cu 25% + Al$_2$O$_3$ 75% seed layer.

Figure 5 Scratch test result of Cu 25% + Al$_2$O$_3$ 75% seed layer

This scratch test is performed by steel wire and scratched seed layer directly. From this figure, Cu 25% + Al_2O_3 75% seed layer was injured by test, but did not peel off. However, Cu 50% + Al_2O_3 50% layer or Cu 100% layer were destroyed by scratch test like Figure 6.

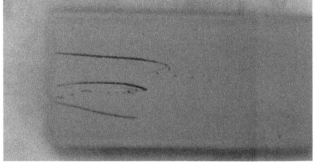

Figure 6 Typical example of destroyed seed layer by scratch test

From these results, we judged Cu 25% + Al_2O_3 75% seed layer is enough tough and low resistance layer for plating. Therefore plating experiment is performed on Cu 25% + Al_2O_3 75% seed layer

Next, we tried to perform electrical plating on Cu 25% + Al_2O_3 75% seed layer. Plating liquid was CuSO4, and anode was Cu. Figure 7 shows typical result of electrical plating.

Figure 7 Typical result of Cu-plating on seed layer

From this figure, fine plating was performed on seed layer, and seed layer did not dissolve or disappear during plating process. This result shows seed layer enough tough to endure from chemical attack of plating liquid. And this result also indicates seed layer can keep high conductivity during plating process. Thus we found our seed layer has enough spec to keep plating conditions, though Cu-Al_2O_3 AD film is made by impact energy between particle and substrate in room temperature.

After plating, we investigated plating layer properties. The thickness of Cu-plating is about 10 μm, and we cannot find peeling of plating layer from seed layer. And we tried to check Vickers hardness to investigate mechanical property. The hardness of plating layer was between 90.9 and 114.5. This value is same as typical Cu film made by plating. Therefore, our plating on seed layer is same as fine Cu plating. Furthermore, to check the plating layer adhesion, we grinded and polished using sand paper and polishing goods. The plating layer became smooth like Figure 8, but plating layer remained without peeling. From this result, good adhesion of plating layer is proved from this result.

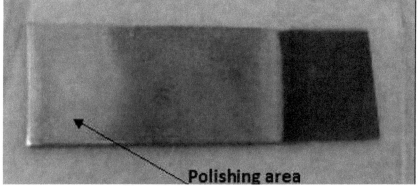

Figure 8 Polishing result of plating layer

And we tried to check specific electrical resistance to investigate electrical property of plating layer. The specific electrical resistance of plating layer was $1.9 \times 10^{-6} \Omega \cdot cm$. This value is almost same to bulk Cu, and this result shows fine Cu plating was performed on insulator by making seed layer using ADM.

In addition, we tried to perform Cu plating on Si substrate by making same seed layer and performing same plating of this glass plating. The result is that we can make fine and tough plating on Si substrate too.

CONCLUSION

In this study, we can perform thick and good electrical plating on glass by making tough and high conductivity seed layer using AD process. Main results are follows.

We can make tough and high conductivity seed layer using ADM. The seed layer is enough tough, low conductivity, chemical stability to perform electrical plating. This characteristic of seed layer can apply power electronics etc.

And we can perform electrical plating on insulator by making seed layer before plating. And the plating is enough tough to endure hard polishing

REFERENCES

1) Jun Akedo, "Aerosol Deposition of Ceramic Thick Films at Room Temperature: Densification Mechanism of Ceramic Layers", Journal of the American Ceramic Society, Vol. 89, No. 6, p.1834–1839, June, 2006

2) Jun Akedo, "Room Temperature Impact Consolidation (RTIC) of Fine Ceramic Powder by Aerosol Deposition Method and Applications to Microdevices", Journal of Thermal Spray Technology, Vol. 17, No. 2, p.181-198, 2008

FORMATION AND ELECTROMAGNETIC PROPERTIES OF 0.1BTO/0.9NZFO CERAMIC COMPOSITE WITH HIGH DENSITY PREPARED BY THREE-STEP SINTERING METHOD

Bin Xiao, Juncong Wang, Ning Ma, Piyi Du *

State Key Laboratory of Silicon Materials
Department of Materials Science and Engineering, Zhejiang University
Hangzhou, 310027, China

*Corresponding Author,   Email address: dupy@zju.edu.cn

ABSTRACT

$BaTiO_3/Ni_{0.5}Zn_{0.5}Fe_2O_4$ (BTO/NZFO) ceramic composite exhibit giant permittivity and high permeability simultaneously and could be potentially used in electronic devices. In this work, the sol-gel *in situ* derived 0.1BTO/0.9NZFO ceramic composites were sintered by conventional ceramic method and a modified three-step sintering process. In the new method, the ceramic composite was sintered initially at 1310 °C for 12 h, then sintered at 1325 °C for only 5 min, and ultimately cooled down rapidly within about 10 seconds. XRD, SEM and EDS were performed to analyze the phase composition, observe the morphology, and identify the elemental distribution of the ceramic composite. Results showed that the composite can be successfully formed through a carefully controlled step-sintering and rapid-cooling process. The composite exhibits a denser microstructure with larger grain size and increased density (by about 10%) compared with those prepared by conventional ceramic method. The effective permittivity of the newly-prepared composite is larger than 270k below 100 kHz, 3.5 times larger than that conventionally sintered at 1310 °C. A relatively low loss (~0.3) was obtained at 30 kHz, 10.6% lower than conventionally prepared samples. The effective permeability reaches ~150 (increased by 16.8%) with low magnetic loss (~0.02). The three-step sintering method is a new effective way to break through the limitation of conventional sintering method and prepare ceramic composites with high density and thus enhanced dielectric and magnetic properties.

INTRODUCTION

Multi-functional materials, such as single-phased multiferroic materials or ferroelectric/ferromagnetic composites, are of great interest to scientists because they are expected to meet ever-growing requirements in advanced electronic devices.[1-5] Ferroelectric/ferromagnetic composites combining two or more functional properties in one entity are potential candidates to realize the purpose of miniaturization and multi-functionality in a convenient way, thus much efforts have been devoted to the exploration of assorted ferroelectric/ferromagnetic composites and investigation of their performance.[6-8] Percolative ferroelectric/ferromagnetic composites, which are attractive to materials researchers recent years because of their simultaneous possession of electrical/magnetic responses, are especially sensitive to preparation details.[9-12] As is known, different synthesis methods customarily lead to different properties of the target materials. With regard to ceramic composites, the influential factors usually include preparation method, starting materials, specific sintering technology and the final structural details that may ultimately endow the composite with prolific dielectric and magnetic properties.[13-16]

BTO/NZFO is a typical percolative ceramic composite exhibiting both excellent dielectric and magnetic properties under externally applied fields. In conventional sintering process, the density of the BTO/NZFO ceramic composite will decrease when the sintering temperature is overly high; hence it is difficult to prepare target ceramic composite with further enhanced functional properties. Obviously, one cannot obtain well-crystallized dense composites by simply further increasing the sintering temperature, because many defects may occur under such conditions. During traditional thermal treatment of ceramic calcination, especially at high temperatures near the melting point, the liquid phase occurring in the system may generate pores due to the mismatch between the existed solid phase and emerging amorphous phase, damaging the electromagnetic properties of obtained composite. However, if the amount of such liquid phase is carefully controlled, it may not cause detriment to the densification of the composite but reversely improve the growth of crystalline grains, resulting in high density. In the initial stage of ceramic calcination when liquid phase start to form, the crystalline grains grow fast. If the sample were rapidly cooled down before plenty of pores begin to form, the liquid phase would be retained, promoting the densification of the composite. It is a possible way to prepare dense composite by carefully controlling the duration around the critical temperature that engenders liquid phase in the system.

Since the appropriate sintering range of BTO/NZFO is found to lie between 1280 °C-1310 °C,[17] and phenomenally, the melting point of BTO/NZFO is found to locate just around 1325 °C. It has been reported that BTONZFO ceramic composite with 90% ferrite content shows satisfactory properties such as giant permittivity and high permeability,[17,18] On the basis of aforementioned idea, we explored a new approach herein to prepare 0.1BTO/0.9NZFO ceramic composite with high density through a modified three-step sintering method. By this method, we successfully obtained dense 0.1BTO/0.9NZFO ceramic composite with improved dielectric and magnetic properties. We believe this work may provide some useful conceptions for obtaining high-property ceramic composite for applications in related devices.

EXPERIMENTAL

The composition of the ceramic composite was fixed as $0.1BaTiO_3$ (BTO)/$0.9Ni_{0.5}Zn_{0.5}Fe_2O_4$ (NZFO) and prepared by a sol-gel in situ process through the following procedures. In the first step, stoichiometric amount of barium acetate ($Ba(CH_3COO)_2$) and nickel acetate ($Ni(CH_3COO)_2 \cdot 4H_2O$) were used as starting materials to synthesize the first sol precursor. They were dissolved in acetic acid (CH_3COOH, 100ml) and stirred for about 60 minutes until a green transparent sol was obtained. For the preparation of the second sol precursor, iron nitrate ($Fe(NO_3)_3 \cdot 9H_2O$), zinc nitrate ($Zn(CH_3COO)_2 \cdot 4H_2O$) and tetrabutyl titanate ($Ti(OC_4H_9)_4$) were used as raw materials and ethylene glycol monomethyl ether ($CH_3OCH_2CH_2OH$, 50~100ml) were used as solvent. The two sol precursors were mixed by adding sufficient deionized water and stirred for about 60 minutes to obtain the final sol. The sol was dried at 120 °C for 2~3 days in oven and gradually turn into composite powder. After being pre-sintered at 750 °C for 1.5h, the powder was added 5% PVA and ground for sufficiently long time before it was pressed into toroidal rings (200 MPa, non-axial). The mass (m) and geometric parameters including the outer diameter (D), inner diameter (d) and height (h) of the samples were measured in order to calculate the relative density (ρ) by the following formula: $\rho=m/V$, in which the volume $V=\pi h(D^2-d^2)/4$. The green bodies are approximately 20 mm in outer diameter, 10 mm in inner diameter and 1-2 mm in height. During the final sintering process, two different sintering methods were applied with the same heating rate (5 °C /min). In one method, the samples were sintered by conventional ceramic process within the temperature range of 1050 °C~1280 °C (in air, 12 h), and then naturally cooled down to room temperature in furnace. In the modified method, the samples were sintered using a three-step way: they were sintered initially at 1310 °C for 12 h, then sintered at 1325 °C for only 5min, and ultimately cooled down from 1325 °C to room temperature very fast outside furnace within about 10 seconds.

The phase composition of as-prepared samples was analyzed by RIGAKUD/MAX-C type X-ray diffractometer (XRD, Cu Kα, λ=0.1540562 nm) between 10°-80°. The morphology was observed by SIRION-type field emitting scanning electron microscope (SEM, produced by Japan FEI Corporation), connected to which is an energy dispersive spectrometer (Oxford, UK) that was used to conduct energy dispersion spectroscopy (EDS). After silver electrodes were painted onto the samples, the conductivity, effective permittivity and dielectric loss of the composites were measured using Agilent 4292A precision impedance analyzer (HP4294A-LRC) under an applied frequency range between 40 Hz and 15 MHz. The effective permeability and magnetic loss were measured by Agilent 16451B precision impedance analyzer (Palo Alto, CA) in the range of 100 kHz-110 MHz.

RESULTS AND DISCUSSION

Figure 1 shows the XRD patterns of the 0.1BTO/0.9NZFO ceramic composite prepared under different sintering conditions. It can be seen that two sets of diffraction peaks could be observed for the samples sintered at 1050°C (S1050), 1150°C (S1150), 1200 °C (S1200), 1280°C (S1280) and 1310 °C (S1310) for 12h, respectively. These diffraction peaks indicate the existence of BTO phase and NZFO phase without any impure phases in the composites. Although the molar content of BTO is only 10%, its assigned peaks could still be identified clearly for the samples sintered under 1280 °C. For Sample ST which was sintered by a modified three-step sintering method (at 1310 °C for 12h, 1325 °C for 5min, then rapidly cooled), the diffraction peaks of the BTO phase could hardly be observed.

The detailed area on the 28-36 degree section is zoomed in for clarity. It is seen that the peaks of the BTO phase gradually diminish with elevated sintering temperature, while the peak positions of the NZFO phase shift toward small angles first below 1280 °C, then move toward large angles above 1280 °C. The peak shift of the NZFO phase below 1280 °C suggests the mutually substitution effect of Ti^{4+} for Ni^{2+}, Zn^{2+} and Fe^{3+} in spinel lattice, but above 1280 °C the lattice parameter of NZFO may decrease due to volatilization of zinc and generation of oxygen vancacies. [17]

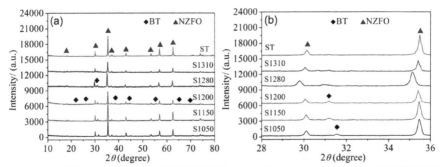

Fig. 1 (a) XRD patterns of the 0.1BTO/0.9NZFO composite ceramics sintered at 1050°C, 1150°C, 1200 °C, 1280°C and 1310°C for 12h, respectively. Sample ST was prepared by a modified three-step sintering method, in which it was sintered initially at 1310 °C for 12 h and sintered continuously at 1325 °C for 5min, then rapidly cooled down outside furnace within about 10 seconds. (b) The detailed area zoomed in on the 28-36 degree section.

The SEM images of the fractures of the 0.1BTO/0.9NZFO ceramic composite prepared under different sintering conditions are shown in Figure 2. As is clearly demonstrated, the crystallinity and the grain size of the composite increase remarkably when the sintering temperature increases from 1050°C to 1310°C as shown in Fig. 2(a)-(e). The samples sintered at low temperatures exhibit a porous structure with small-sized crystalline grains and disconnected topological microstructure. The grain size of the composite becomes much larger and the density of the sample increases with elevated sintering temperature. Perfect crystalline phase with large size appears, but still not so dense enough in the samples sintered at a high temperature such as 1280 °C as shown in Fig. 3(d). The densest structure with largest grain size is observed in Sample ST that was prepared by modified three-step sintering method as shown in Fig. 3(f).

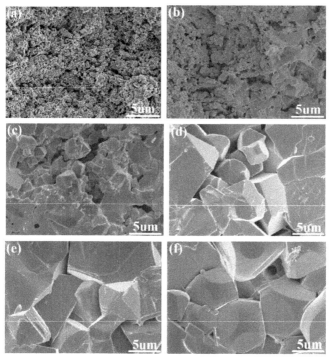

Fig. 2 SEM images of the fractures of the 0.9NZFO/0.1BT composite ceramics: (a) S1050; (b) S1150; (c) S1200; (d) S1280; (e) S1310; (f) ST.

The plots of the density of the 0.1BTO/0.9NZFO ceramic composite sintered at different conditions are calculated in Figure 3. It is seen that elevated sintering temperature could result in higher density of the composite from 2.90 g/cm^3 (1050 °C) to 4.55 g/cm^3 (1310 °C). The highest density of 5.00 g/cm^3 appears in Sample ST which experienced a modified three-step sintering method and rapidly cooled down from 1325 °C to room temperature within about 10 seconds.

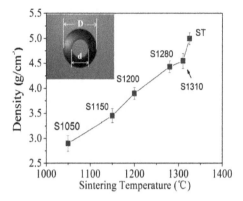

Fig. 3 Plots of the density of the 0.1BTO/0.9NZFO composite ceramics sintered at different conditions. The inset shows a photo of the ceramic sample. The density (ρ) is calculated by the following formula: ρ=m/V, in which m is the mass, V=πh(D^2-d^2)/4 is the volume with h being the height (thickness), D the outer diameter, d the inner diameter of the toroidal sample.

The energy dispersion spectroscopy (EDS) conducted on a selected crystal area in Sample ST prepared by three-step sintering method is illustrated in Figure 4. The distribution of the constituent elements clearly shows the phase composition of the sample. It is seen that the agglomeration of Ba and Ti element constituting the BTO phase appear mainly between the interfaces of the NZFO particles, existing in the shape of thin layers in the composite with much smaller grain size.

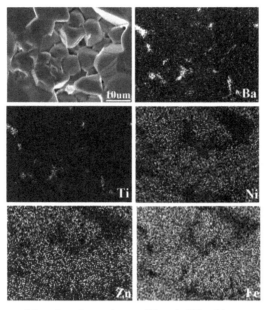

Fig. 4 SEM image of the selected crystal area of Sample ST and its energy dispersion spectroscopy (EDS) results

Obviously, for the samples sintered under different conditions as shown in Fig. 1, the crystalline NZFO phase could easily form in the ceramic composite. However, the BTO phase only formed in the ceramics prepared by conventional sintering method, and almost no diffraction peaks of it could be obsrved in the composite prepared by three-step sintering method. Moreover, the 0.1BTO/0.9NZFO ceramic composite cannot form dense structure when the sintering temperature is lower than 1280°C as shown in Fig. 2, which is believed to fall out its most appropriate sintering range.[17]

Generally, there are two factors influencing the density of ceramics. On one hand, at low temperatures below 1280 °C in this work, the crystalline phase forms not so perfect that the ceramics are not well calcined and have typically porous structure inside. Therefore, the grain size of the crystalline phase and the density of the ceramics increase with elevated sintering temperature as shown in Fig. 2(a)-(d) and Fig. 3. The higher the temperature is, the higher the density of ceramics will be. On the other hand, at excessively high sintering temperature, over-sintering of ceramics would occur, deteriorating the dense microstructure of the composite by reversely generating big pores and more defects that may come from oxygen vacancies, migration of grain boundaries, mismatch between heterogeneous particles and rapid volume shrinkage during the calcination process in the composite system.[19] Hence, further increasing the sintering temperature would cause a considerable decrease in the density of the composite. Practically, the appropriate temperature for the calcination of BTO/NZFO composite seems to locate between 1280 °C and 1310 °C with the density being 4.43 g/cm^3 and 4.55 g/cm^3, respectively.

In addition, it is known that liquid phase with high activation for ionic migration may promote the growth and crystallization of the constituent phases. In this work, it was found that the melting point of sol-gel derived BTO/NZFO composite is about 1325°C, only 15 °C higher than the upper limit of the appropriate sintering range. Therefore, elevating the sintering temperature to 1325°C after the sample has been calcined at 1310 °C for 12h and preserving for a very short time may promote the crystallization and increase further the density of the composite becasue highly activated liquid phase appears possibly along the grain boundaries. The preserving time at 1325°C was precisely controlled to be only 5min, which allowed the liquid phase to form and also too short for the composite to melt since melting is a dynamic process that requires sufficient energy and time for the constituent elements to diffuse and migrate. As shown in Fig. 2(f) and Fig. 3, the grain size and the density of as-prepared sample (ST, rapid cooled) are apparently enhanced compared with that sintered at 1310 °C. Furthermore, the existence of amorphous BTO phase could be evidenced by XRD pattern and EDS results as shown in Fig. 1 and 4. It can be seen that the elements of Ba and Ti exist among the interfaces of NZFO particles but the BTO crystalline phase does not appear in the XRD pattern of Sample ST. It is evident by EDS that the BTO phase may melt into amorphous state when the sample was sintered at an extremely high temperature like 1325 °C, which is just in the vicinity of the melting point of the composite. Such amorphous state could be successfully retained during a rapid cooling process, because the cooling speed is too fast to allow the amorphous-phased BTO phase to re-crystallize. Hence, the modified three-step sintering method is obviously an effective way to increase the density of the composite. The density of the 0.1BTO/0.9NZFO composite is raised to 5.00 g/cm^3, which is 10.5% higher than that calcined by conventional ceramic method.

Figure 5 shows the variation of effective permeability and magnetic loss of the composite versus externally applied frequency. For clarity, the initial permeability of the composite as a function of sintering temperature is depicted in Fig. 5(c). The initial permeability of S1280 and S1310 could reach ~125 and ~128, respectively, but that of Sample ST has achieved ~150, increased by 16.8% compared with that calcined by conventional method at 1300 °C. Also, the low magnetic loss of Sample ST is only ~0.02, covering a wide frequency range up to several megahertz. The improved magnetic properties of Sample ST are due to the high density as well as large grain size as mentioned above.

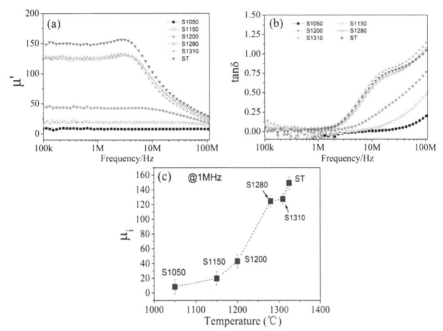

Fig. 5 The frequency dependence of (a) effective permeability (b) magnetic loss of the 0.1BTO/0.9NZFO composite ceramics. (c) Plots of initial permeability as a function of sintering temperature.

Figure 6 shows the frequency dependence of the electrical conductivity of the composite and the dc conductivity in variation with sintering temperature. As seen, the dc conductivity of the composite is as small as $\sim 10^{-4}$ S/m for the samples with loose microstructure and substantial pores, however, the conductivity achieves as high as almost $\sim 5 \times 10^{-2}$ S/m, about two orders of magnitude higher with the improvement of sintering conditions. Note that Sample ST has the largest dc conductivity (4.928×10^{-2} S/m), which is in accordance with the dense morphology as observed in Fig. 2(f). Possessing dense microstructure with high density, the dispersion frequency of Sample ST is also the highest, well reflecting the effects of decreased resistivity, increased grain size, and better crystallization of the conductive NZFO phase.[20]

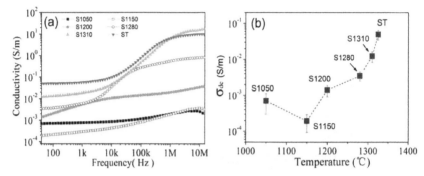

Fig. 6 (a) The frequency dependence of the electrical conductivity of the 0.9NZFO/0.1BT composite ceramics; (b) The dc conductivity of the composite sintered at various temperatures.

The frequency dependence of the effective permittivity and dielectric loss of the 0.1BTO/0.9NZFO composite ceramics is shown in Figure 7. The effective permittivity of the newly-prepared composite has been raised to >270k (at 100 kHz), 3.5 times larger than that conventionally sintered at 1310 °C. Furthermore, the loss peaks of the samples as shown in Fig. 7(b) are found to shift toward higher frequency as the sintering temperature increases. The lowest dielectric loss is impressively as low as ~0.304 (at about 30 kHz), covering a wide frequency range from 1kHz to 1 MHz for Sample ST that was prepared by modified three-step sintering method, 10.6% lower than conventionally sintered sample (~0.341) and 41.5% lower than that reported in literature. [18]

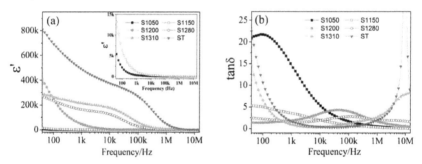

Fig. 7 The frequency dependence of (a) effective permittivity (b) dielectric loss of the 0.1BTO/0.9NZFO composite ceramics sintered under different conditions

In a word, the modified three-step sintering method is an effective way to prepare ferroelectric/ferromagnetic composite ceramics with high density and enhanced dielectric and magnetic properties.

CONCLUSIONS

A modified three-step sintering method was proposed for the preparation of 0.1BTO/0.9NZFO ceramic composite derived from sol-gel *in situ* process. Compared with conventional sintering method, the new approach could successfully obtain 0.1BTO/0.9NZFO composite with high density, which is raised to 5.00 g/cm^3, 10.5% higher than that calcined by conventional ceramic method. Such high density is mainly attributed to the formation of amorphous BTO phase during the three-step sintering process. Resultantly, the as-prepared composite exhibited enhanced dielectric and magnetic properties to a considerable extent. It exhibited giant effective permittivity (>270k till 100 kHz) that is 3.5 times larger than that of conventionally sintered samples, meanwhile, a minimum dielectric loss of ~0.304 was obtained at 30 kHz, 10.6% lower than conventionally sintered samples at 1310 °C (~0.341). The effective permeability of the newly-prepared composite could reach ~150 (increased by 16.8%) and the magnetic loss is only ~0.02 below 1 MHz. The modified three-step sintering method is an effective way to obtain ceramic composite with high density as well as excellent dielectric and magnetic properties.

ACKNOWLEDGEMENTS

This work was supported by the Natural Science Foundation of China under grant No. 51272230 and No. 50872120, Zhejiang Provincial Natural Science Foundation (Grant No. Z4110040) and the National Key Scientific and Technological Project of China (Grant No. 2009CB623302).

REFERENCES

[1]L. He, D. Zhou, H. Yang, J. Guo, and H. Wang, A Novel Magneto-Dielectric Solid Solution Ceramic 0.25LiFe$_5$O$_8$–0.75Li$_2$ZnTi$_3$O$_8$ with Relatively High Permeability and Ultra-Low Dielectric Loss, *J. Am. Ceram. Soc.*, **95**, 3732-4(2012).

[2]I. Jung and J. Y. Son, A nonvolatile memory device made of a graphene nanoribbon and multiferroic BiFeO$_3$ gate dielectric layer, *Carbon*, **50**, 3854-8(2012).

[3]J. Q. Huang, P. Y. Du, L. X. Hong, Y. L. Dong, and M. C. Hong, A Percolative Ferromagnetic–Ferroelectric Composite with Significant Dielectric and Magnetic Properties, *Adv. Mater.*, **19**, 437-40(2007).

[4]C. W. Nan, M. I. Bichurin, S. X. Dong, D. Viehland, and G. Srinivasan, Multiferroic magnetoelectric composites: Historical perspective, status, and future directions, *J. Appl. Phys.*, **103**, 031101, 35pp(2008).

[5]T. N. Narayanan, B. P. Mandal, A. K. Tyagi, A. Kumararisi, X. Zhan, M. G. Hahm, M. R. Anantharaman, G. Lawes, and P. M. Ajayan, Hybrid Multiferroic Nanostructure with Magnetic-Dielectric Coupling, *Nano. Lett.*, **12**, 3025-30(2012).

[6]H. Zheng, L. Li, Z. J. Xu, W. J. Weng, G. R. Han, N. Ma, and P. Y. Du, Ferroelectric/ferromagnetic ceramic composite and its hybrid permittivity stemming from hopping charge and conductivity inhomogeneity, *J. Appl. Phys.*, **113**, 044101, 8pp(2013).

[7]A. Sharma, R. K. Kotnala, and N. S. Negi, Structural, dielectric, magnetic and ferroelectric properties of (PbTiO$_3$)$_{0.5}$-(Co$_{0.5}$Zn$_{0.5}$Fe$_2$O$_4$)$_{0.5}$ composite, *Phys. B: Conden. Mater.*, **415**, 97-101(2013).

[8]N. A. Pertsev, S. Prokhorenko, and B. Dkhil, Giant magnetocapacitance of strained ferroelectric-ferromagnetic hybrids, *Phys. Rev. B*, **85**, 134111, 5pp(2012).

[9]Z. R. Wang, T. Hu, X. G. Li, G. R. Han, W. J. Weng, N. Ma, and P. Y. Du, Nano conductive particle dispersed percolative thin film ceramics with high permittivity and high tunability, *Appl. Phys. Lett.*, **100**, 132909, 4pp(2012).

[10]Z. R. Wang, T. Hu, L. W. Tang, N. Ma, C. L. Song, G. R. Han, W. J. Weng, and P. Y. Du, Ag nanoparticle dispersed PbTiO$_3$ percolative composite thin film with high permittivity, *Appl. Phys. Lett.*, **93**, 222901, 3pp(2008).

[11]Z. M. Dang, Y. Shen, and C. W. Nan, Dielectric behavior of three-phase percolative Ni-BaTiO$_3$/polyvinylidene fluoride composites, *Appl. Phys. Lett.*, **81**, 4814-6(2002).

[12]J. W. Xu and C. P. Wong, Low-loss percolative dielectric composite, *Appl. Phys. Lett.*, **87**, 082907, 3pp(2005).

[13]B. Xiao, N. Ma, and P. Y. Du, Percolative NZFO/BTO ceramic composite with magnetism threshold, *J. Mater. Chem. C*, DOI:10.1039/C3TC30807C(2013).

[14]M. A. Ahmed, N. Okasha, and N. G. Imam, Modification of composite ceramics properties via different preparation techniques, *J. Magn. Magn. Mater.*, **324**, 4136-42(2012).

[15]P. S. S. R. Krishnan, Q. M. Ramasse, W. I. Liang, Y. H. Chu, V. Nagarajan, and P. Munroe, Effect of processing kinetics on the structure of ferromagnetic-ferroelectric-ferromagnetic interfaces, *J. Appl. Phys.*, **112**, 104102, 7pp(2012).

[16]G. S. Rohrer, M. Affatigato, M. Backhaus, R. K. Bordia, H. M. Chan, S. Curtarolo, A. Demkov, J. N. Eckstein, K. T. Faber, J. E. Garay, Y. Gogotsi, L. Huang, L. E. Jones, S. V. Kalinin, R. J. Lad, C. G. Levi, J. Levy, J. P. Maria, L. Mattos Jr, A. Navrotsky, N. Orlovskaya, C. Pantano, J. F. Stebbins, T. S. Sudarshan, T. Tani, and K. S. Weil, Challenges in Ceramic Science: A Report from the Workshop on Emerging Research Areas in Ceramic Science, *J. Am. Ceram. Soc.*, **95**, 3699-712(2012).

[17]B. Xiao, Y. L. Dong, N. Ma, and P. Y. Du, Formation of Sol-Gel *In Situ* Derived BTO/NZFO Composite Ceramics with Considerable Dielectric and Magnetic Properties, *J. Am. Ceram. Soc.*, **96**, 1240-7(2013).

[18]H. Zheng, Y. L. Dong, X. Wang, W. J. Weng, G. R. Han, N. Ma, and P. Y. Du, Super High Threshold Percolative Ferroelectric/Ferrimagnetic Composite Ceramics with Outstanding Permittivity and Initial Permeability, *Angew. Chem. Int. Ed.*, **48**, 8927-30(2009).

[19]M. M. Vijatović, J. D. Bobić, and B. D. Stojanović, History and Challenges of Barium Titanate: Part I, *Sci. Sinter.*, **40**, 155-65(2008).

[20]Y. F. Qu, Function Ceramics and Application, pp:250, Edited by Chemical Industry Press, Beijing (2003).

Microwave Materials and their Applications

THIN GLASS CHARACTERIZATION IN THE RADIO FREQUENCY RANGE

Alfred Ebberg[1], Jürgen Meggers[1], Kai Rathjen[1], Gerhard Fotheringham[2], Ivan Ndip[2], Florian Ohnimus[2*], Christian Tschoban[2], Isa Pieper[3], Andreas Kilian[4**], Sebastian Methfessel[4], Martin Letz[5], Ulrich Fotheringham[5]
[1]: West Coast University of Applied Sciences (FHW), Heide, Germany
[2]: Fraunhofer Institute for Reliability and Microintegration (IZM), Berlin, Germany,
[3]: Fraunhofer Institute for Silicon Technology (ISIT), Itzehoe, Germany
[4]: Chair for Microwave Engineering and High-Frequency Technology, Friedrich-Alexander-University Erlangen-Nuremberg (FAU), Germany
[5]: SCHOTT AG, Mainz, Germany
*: Dr. Ohnimus is now with Rohde & Schwarz GmbH, Berlin, Germany.
**: Dr. Kilian is now with Tesat-Spacecom, GmbH & Co. KG, Backnang, Germany.

ABSTRACT
Glass wafers are being widely considered as a transparent and low-permittivity alternative to silicon substrates for radio frequency devices.

To provide reliable material data to the device designer, the complex permittivity of three important wafer glasses with the thermal expansion coefficients matching the one of silicon, namely AF32®eco, Borofloat33®, and MEMpax® from SCHOTT, has been investigated by two different methods (split cylinder and open resonator) in the range from 1 to 100 GHz.

The data sets thus obtained for these glasses show a remarkable consistency, with a relative deviation of 1-2% for the real part of the permittivity and 10-25% for the effective dielectric loss tangent.

In addition to this, the results have been confirmed by an indirect determination of the permittivity from the simulation of special radio frequency devices (resonant microstrip structures on wafers from said glasses as substrates). The relative deviation is about 1-3% for the real part of the permittivity and ca. 50% for the loss tangent this time.

Concerning the effective dielectric losses, the performance of the three glasses can only be matched by (costly) high resistivity silicon rather than by commonly used low resistivity silicon.

INTRODUCTION

Glass wafers are being widely investigated as substrates for radio frequency (RF) devices[1], e.g., RF filters[2], RF antennas[3], and many others. They compete with another "newcomer" in the field of RF substrates, namely silicon itself. Compared to silicon substrates, glass substrates offer a significant reduction of the real part of the permittivity as well as optical transparency, two features that are advantageous for many applications. To obtain a similar compatibility for the RF components to silicon based (other) components, wafer glasses with their thermal expansion coefficients matching that of silicon are preferred as substrate materials.

Due to the development of "through-glass-vias" (TGVs)[4, 5], glass substrates are also equally suited for 3D integration, e.g. by allowing for the replacement of wire connects (bonds) contacting active chip areas from the side of the 3D-stack.

In terms of effective dielectric loss, low-loss glasses exceed the performance of CMOS grade silicon and come close to that of low-loss silicon. As this comparison is the essential motivation for the measurements reported here, the effective dielectric loss of silicon is discussed in the following.

A few remarks on RF structures in general and the role of the real part of the relative permittivity ε_r in microelectronics shall be made first:

1. Signal speed: Transmission lines can be considered as guiding the waves through the substrate. Therefore, the signal velocity depends primarily on the parameters of the substrate, i. e. its permittivity. From this point of view, ε_r should be as small as possible (close to 1), but its precise value is not so important.
2. Embedded capacitances: For miniaturization purposes, capacitances are often realized as embedded devices. For this application, of course should be large, (≥ 20). But this will not be considered in this paper.
3. Resonant structures: The characteristics of resonant structures, e.g. planar antennas, depend sensitively on ε_r. In this case, the designer needs to know its exact value.
4. Miniaturization of antenna structures: For resonant structures like the above mentioned planar antennas, the degree of miniaturization roughly scales with $1/\sqrt{\varepsilon_r}$, making larger values of ε_r favourable. On the other hand, a substrate with a large ε_r tends to "pull" the electric field into its interior, thereby lowering the (antenna) efficiency. This effect could be worsened by a large loss tangent of the substrate. Therefore a trade-off value for ε_r has to be found, which according to the experience of the Fraunhofer IZM group amounts to approximately 5. This is one application where the comparatively small real part of the permittivity of certain glasses is favourable.

This brings us to the discussion of effective dielectric losses, especially those of silicon.

In the low frequency range, the loss tangent of silicon is related to the conductivity and to the real part of the permittivity by[6]:

$$\tan(\delta) = \frac{\sigma}{\varepsilon_0 \cdot \varepsilon_r' \cdot \omega} \tag{1}$$

σ denotes the static conductivity of silicon, ε_r' is the real part of the relative permittivity (value for silicon: 11.9) and ε_0 is the vacuum permittivity. At higher frequencies additional contributions to the loss tangent will be perceptible. This is sometimes written as:

$$\tan(\delta) = \frac{\sigma}{\varepsilon_0 \cdot \varepsilon_r' \cdot \omega} + \tan(\delta)_b \tag{2}$$

The latter term stems from the "bound" electrons. Up to 100 GHz it is comparatively small and can be neglected[7, 8]. Strictly speaking, σ itself is frequency dependent, but again, this effect may be neglected up to about 150 GHz. So, for Si as a substrate material, the loss tangent directly depends on the DC conductivity of the Si wafer. Depending on this conductivity value, silicon may range from a good dielectric to a problematic one. There is a common classification of silicon distinguishing between low, medium, and high resistivity silicon (LRS, MRS, and HRS) with typical values being 1 Ohm·cm, 10 Ohm·cm, and 500 Ohm·cm[9]. The common CMOS grade silicon has a resistivity in the range from 1 to 20 Ohm·cm[10] (or even much less, depending on the degree of doping[11]) so that typically, it compares to LRS, at maximum to MRS.

Employing full wave simulation, Ndip et al. from the Fraunhofer IZM group have found that the transmission loss of through-silicon vias (TSVs) rises sharply in the frequency range 2.4-60 GHz[9]. For this simulation, they considered LRS, MRS, and HRS with the above mentioned typical resistivity values. For 2.4 GHz, they got the following approximate values of power loss through TSV for typical set-ups for WLAN/WPAN applications: 13% for both LRS and MRS, 2% for HRS. For 5 GHz, the corresponding values were: 29% for LRS, 17% for MRS, 2% for HRS. For 60 GHz, the corresponding values were: 76% for LRS, 24% for MRS, 4% for HRS.

For a thorough analysis the semiconducting behaviour of silicon has to be accounted for. For a rough picture, one might use (1) and translate the conductivity of silicon into a loss tangent. For the important frequency range around 24GHz, one gets the approximate values $\tan(\delta)=6.3$ for LRS, $\tan(\delta)=0.63$ for MRS, and $\tan(\delta)=0.013$ for HRS. These figures provide benchmarks for an assessment of glass as substrate material.

Remarkably, glass may provide an alternative to LRS and MRS in terms of effective dielectric losses so that it is an alternative substrate material not only with respect to its optical transparency and small real part of the permittivity, see above. Indeed, there is literature indicating that radio frequency (RF) structures on suited glass substrates show similar losses as the same structures on HRS[12]. Aside from exceptions[6, 7], however, little is known on the complex per-mittivity of glass substrates in the Gigahertz regime which would be necessary for a reliable design of considered RF devices, e.g. planar antennas.

To some extent, this is due to the fact that special experimental efforts are required for that purpose. In the following, a couple of results from such efforts will be presented which yield a consistent picture of the dielectric properties of AF32®eco, Borofloat33® □and MEMpax® from SCHOTT. AF32®eco is an alkaline-free alumosilicate glass produced by drawing. Boro-float33® □and MEMpax® are both borosilicate glasses with similar chemical and physical properties. The former is produced by floating, the latter by drawing. All three are important optoelectronic substrate glasses, with the thermal expansion coefficient matching the one of silicon (in contrast to vitreous silica, for instance).

Same as in[6], two essentially different measurement types are used, i.e. resonator methods where the impact of the insertion of a large piece of material on the quality factor of the resonator is measured and waveguide methods where the impact of the substrate material properties on the quality factor of a resonating waveguide structure carried by this glass substrate is derived from a comparison of measurement results with corresponding simulations.

In contrast to[6], however, the resonating waveguide structures are not coplanar ones but of the microstrip type where the electromagnetic power fraction concentrated in the substrate is considerably higher.

The resonator methods have the advantage of being direct measurements of material pro-perties. The waveguide methods have the advantage of being closer to the set-up of the products where the material properties shall enter the design process. However, it is quite a challenge to separate the losses in the dielectric from the losses in the conductor and from radiation losses which is necessary with the waveguide methods. Obviously, carrying out both types of measure-ments and finding the results to coincide provides the optimum basis for product design.

GLASS CHARACTERIZATION BY RESONATOR METHODS

The most accurate methods to determine ε' and $\tan(\delta)$ of a material in the microwave region are resonator methods which use a large piece of material (with the preferred minimum dimension equalling half the wavelength in the material), e.g. a solid cylinder or a block in a hollow wave guide in their resonating structure.

Since we are interested in the characterization of materials shaped as thin wafers, however, we need to choose methods which allow the determination of material properties from thin glass sheets, i.e. plate like geometries. Split-cylinder cavities with a high quality metallization are well suited for such dimensions and results using this method are shown in the next section. Open resonator measurements on thin glass sheets are shown in the over next section.

By split cylinder resonator

For a few years, a split-cylinder resonator has been commercially available from Agilent which has been used for the glass characterization carried out at Fraunhofer IZM. The two halves of the cylinder can be separated, and the wafer under test may be inserted in between, allowing for non-destructive analysis. Because of the split, there is some radiation loss, which is accounted for by the evaluation routine, based on[13]. For evaluation, the TE_{0np} resonances have to be used. Because of the radial symmetry (subscript 0), only circular surface currents are induced in the walls of the cylinder, so that the overall field pattern is not affected so much by the "split". Moreover, for these resonances the transport of energy into the walls is relatively small. "n" may be arbitrary, but "p" has to be odd. This guarantees an anti-node in the middle of the cylinder, so that there is maximum interaction between the electric field and the wafer (Fig. 1).

Simulated electric field magnitude (red→high, green→low)

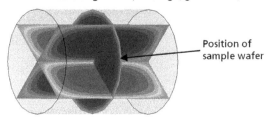

Position of sample wafer

Figure 1: 3D simulation model of the split cylinder resonator showing the computed electric field pattern of the TE_{011} cavity resonance mode (from IZM, using ANSYS HFSS®)

At higher frequencies, however, the usable frequencies are not easily detectable and separable from the myriad of resonances, so that the method is not as straight forward as it may seem. Moreover, the method heavily relies on an exact knowledge of the wafer's thickness and homogeneity, so that great care was taken to determine both. IZM has been able to use this method up to 30 GHz, in one case up to 40 GHz[14]. The uncertainty has been estimated to be 2-3 % for the real part of the relative permittivity and 0.001 for the loss tangent.

In the following, the IZM results for AF32®eco and Borofloat33® (4 inch wafers, 0.5mm thickness) are presented (Fig. 2).

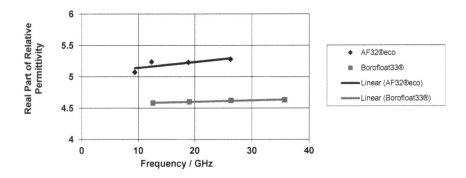

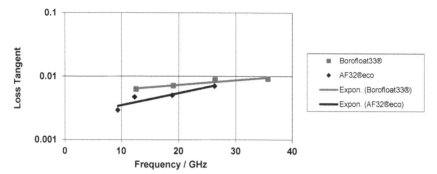

Figure 2: Relative permittivity of AF32®eco and Borofloat33® in the GHz regime, as measured by IZM. As an aid to the eye, linear interpolations of the real parts of the permittivity data and exponential interpolations of the loss tangent data have been added.

By open resonator

At the FAU, the materials have been measured in a hemispherical open resonator setup at 24GHz and 77GHz. As shown in fig. 3, these Fabry-Perot interferometers generate a focussed beam between a concave and a plane mirror with a Gaussian profile of the electric field in radial direction. The resonators are excited by rectangular wave-guides which are terminated by coupling apertures in the surface of the concave mirror.

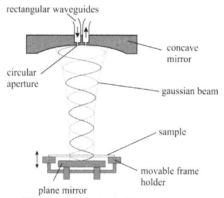

Figure 3: Hemispherical Open Resonator

According to the radius of curvature and the distance between the mirrors a standing wave of the fundamental transversal electromagnetic mode (TEM_{00n}) is excited at a defined resonance frequency. A vector network analyzer is used to measure this frequency and the quality factor of the resonance curve, resp. Then the sample is inserted into the beam to quantify the detuning of these two parameters. By this perturbation method, the complex permittivity can be derived. At best, the specimen thickness should equal a multiple of half a wavelength in the material. This guarantees lowest uncertainties due to the phase front corrections made in the theory and analysis of the open resonator. As the thickness of the characterized samples was lower than half a wavelength a modified procedure has been applied[15]. The sample is placed onto a movable frame holder right above the plane mirror into the waist of the beam. The measurement procedure starts by searching the sample position of highest sensitivity where the antinode of the electric field in longitudinal direction coincides with the centre of the specimen. By use of a multi-layer solution for the open resonator the material properties can be calculated. The worst-case uncertainty of this setup has been estimated deterministically as ± 0.5 % for the real part of the relative permittivity and ± 18 % for the loss tangent. In fig. 4, the measurements at 4 inch wafers from AF32®eco, Borofloat33® (both 1.1mm thick) and MEMpax® (0.5mm) are presented.

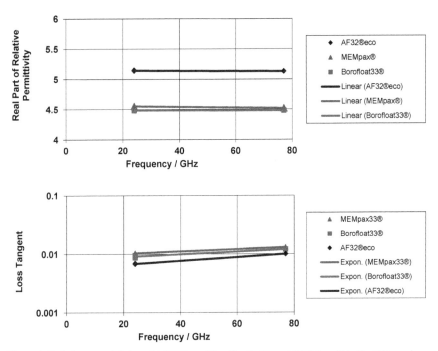

Figure 4: Relative permittivity of AF32®eco, Borofloat33®, and MEMpax® in the GHz regime, as measured by FAU. As an aid to the eye, linear or exponential interpolations have been added.

In addition, the FAU has characterized AF32®eco, Borofloat33®, and MEMpax® in the 150-210 GHz range by a non-resonant, quasi-optical, mm-wave radar setup[16] yielding the following mean values for the dielectric properties:

Glass	$\varepsilon_r{}'$	$\tan(\delta)$
AF32®eco	5.03	0.0154
Borofloat33®	4.38	0.0161
MEMpax®	4.4	0.0235

Table 1: Mean values of the relative permittivity of AF32®eco, Borofloat33®, and MEMpax® in the 150-210GHz regime, as measured by FAU.

Note that the $\varepsilon_r{}'$ and $\tan(\delta)$ values derived by the latter type of measurement oscillate over the frequency range investigated, with an amplitude of 0.05 for $\varepsilon_r{}'$ and of 0.08 for $\tan(\delta)$, so that these values have to be considered as rough estimates.

Combining all results in the 4GHz-77GHz range yields (fig. 5):

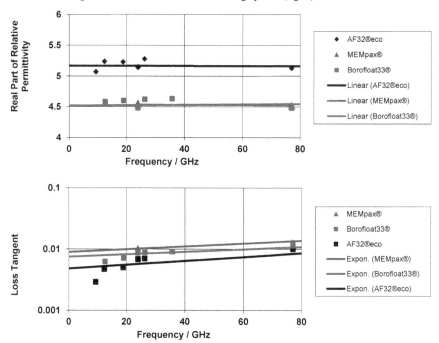

Fig. 5: Compilation of the relative permittivities of AF32®eco, Borofloat33®, □and MEMpax®□ in the GHz regime. As an aid to the eye, linear or exponential, resp., interpolations have been added. Note that for the interpolation of the loss tangent, one data pair from the measurement in the 150-210GHz range has been added (180GHz, mean value).

GLASS CHARACTERIZATION BY WAVEGUIDE METHODS

As already stated, the use of microwave waveguides allows the measurement of $\varepsilon_r{}'$ and $\tan(\delta)$ in a configuration very similar to the setup of the final products, where the measured parameters will be used in the design process. Microstrip and coplanar waveguides can be easily fabricated on glass substrates using standard technologies from semi-conductor industry. Relative permittivity and losses of the substrate strongly determine transmission and/or reflection properties of the guiding structures, and can therefore be determined from appropriate measurements. Fig. 6 shows the distribution of the electrical field of the principal waveguide modes in a microstrip and a coplanar waveguide simulated with CST MICROWAVE STUDIO®.

As can be seen from the figures the electrical field is much more concentrated within the substrate in the case of the microstrip waveguide, compared to the coplanar waveguide. Therefore the sensitivity on variations of $\tan(\delta)$ are expected to be higher using microstrip lines.

a. Coplanar waveguide   b. Microstrip waveguide
Figure 6: Electrical field distribution of a coplanar and a microstrip waveguide

Ring resonators are commonly used to determine the properties of microwave substrates, in particular ε_r' and $\tan(\delta)$. A ring resonator is made from a closed loop of an either coplanar or microstrip transmission line. The resonator acts as a band-pass filter, exhibiting low insertion loss at frequencies, where the mean circumference of the ring is equal to an integral multiple of the guided wavelength. Ring resonators offer low radiation losses and high Q values. Coupling to the resonator can be performed by two gaps, designed for loose coupling (fig. 7).

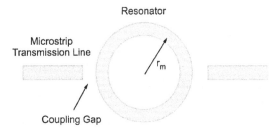

Figure 7: Ring resonator

Using a full 3D electromagnetic simulation with CST MICROWAVE STUDIO®, the insertion loss as a function of $\tan(\delta)$ for ring resonators realized in microstrip and coplanar technology have been calculated. The results are shown in fig. 8. For comparison the insertion losses for straight waveguides have been included.

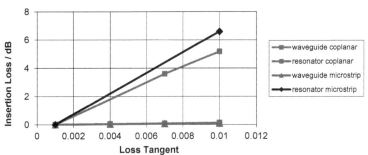

Figure 8: Insertion loss as function of loss tangent for different waveguide structures

As can be seen from fig. 8, a microstrip resonator offers the maximum slope and therefore the best sensitivity to measure low losses in glass substrates.

One problem to overcome, when using microstrip lines, is to contact ground and signal with the measurement system, usually a microwave network analyzer. Commercially available probes usually use a ground signal ground (GSG) configuration, which is very well suited for contacting coplanar waveguide structures. For contacting microstrip structures, ground from the bottom of the substrate has to be transformed to the top layer. This can either be done by a via or by using microstrip radial stubs[17]. Fig. 9 shows the simulation model of a micro-strip ring resonator with a radial stub designed to contact a microstrip transmission line with a GSG probe.

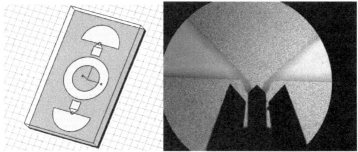

Figure 9: Radial stub designed to contact a microstrip line with a GSG probe and photograph

Microstrip resonators of different diameters suited for measurements in a frequency range from 2 GHz to 45 GHz as well as contacting and deembedding structures have been fabricated on glass wafers by Fraunhofer ISIT. Measurements of the transmission characteristics of the resonators have been performed using a microwave network analyzer. The glass wafers were Borofloat33®, AF32®eco, and vitreous silica (Lithosil®) provided by SCHOTT (6 inch wafers, 0.675mm thickness). The photograph of a wafer is shown in fig. 10.

Figure 10: Glass wafer with microstrip ring resonators

Fig. 11 shows the measured insertion losses of 3 GHz resonators on different substrates. The resonant frequency, the insertion loss at resonance and the bandwidth of the transmission characteristic strongly depend on the substrate's relative permittivity and loss factor.

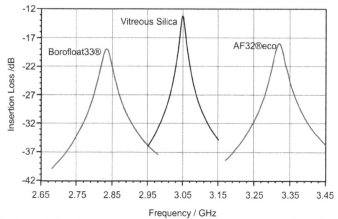

Figure 11: Transmission characteristics of 3 GHz resonators on different glass substrates. The shape of the resonance curves varies with frequency, as well as the insertion loss at resonance.

By fitting these measurement results with the results of a full 3D field simulation of the setup, the real part of the relative permittivity ε_r' and loss factor $\tan(\delta)$ of the substrates can be determined. The 3D simulation takes into account radiation losses and losses due to skin effect as well.

Fig. 12 shows the values of ε_r' and $\tan(\delta)$ for different glass substrates as a function of frequency, obtained from measured and simulated transmission characteristics of ring resonators. Note that the $\tan(\delta)$ values for vitreous silica are high compared to, for instance, the findings of the IZM group who had not been able to determine those values at a satisfactory precision because they stayed below 0.0006 throughout the frequency range investigated. The values found here may indicate the limits of the waveguide method which implies the calculation of small differences of big quantities (measured loss minus losses from other loss mechanisms).

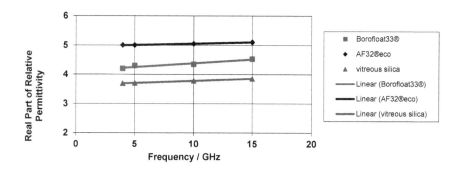

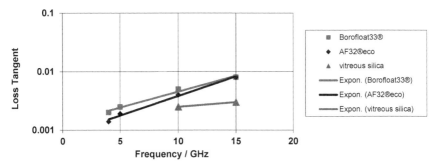

Figure 12: Relative permittivity of AF32®, Borofloat33®, and vitreous silica in the GHz regime, as measured by FHW. As an aid to the eye, linear or exponential interpolations have been added.

DISCUSSION OF RESULTS

Considering both the real part of the relative permittivity, ε_r', and the loss tangent, $\tan(\delta)$, of all three glasses in the range from 4GHz to 77GHz, one finds almost constancy for ε_r' and an increase by less than half an order of magnitude for $\tan(\delta)$, see fig. 13.

The mean values of ε_r' and the root mean square (RMS) relative deviations from the latter are listed in table 2. The order of magnitude of the relative deviations is 1-2% despite the involvement of two different measurement set-ups.

Remarkably, the values obtained from the waveguide methods fit well to the values obtained from the resonator methods. Considering the RMS relative deviations of ε_r' derived from waveguide data to the constant values from above, one finds 3-5% difference, see again table 2.

No relative deviations are given for MEMpax® because of the few measurements made with respect to the chemical and physical similarity to Borofloat33®. The difference in complex permittivity of them is not significant anyhow in terms of the measurement reproducibilities.

Glass	Mean value of resonator measured ε_r'	RMS relative deviation of resonator-measured ε_r' from mean value	RMS relative deviation of wave-guide-determined ε_r' from resonator-determined mean value
AF32®eco	5.1658	1.1%	2.6%
Borofloat33®	4.5176	1.4%	4.7%
MEMpax®	4.535	-	-

Table 2: Mean values and RMS deviations of ε_r' in the frequency range 4GHz-77GHz for AF32®eco, Borofloat33® and MEMpax® determined by resonator and waveguide methods.

As for $\tan(\delta)$, one finds the RMS relative deviations of the data from an exponential fit to amount to 10-25%, see table 3 and fig. 13. This indicates a remarkable consistency of the resonator measurements involved. (Note that the reproducibility of the individual methods has the same order of magnitude concerning the loss angle.) Concerning the consistency of the resonator-measured data with the waveguide-determined ones, there is a rough, but significant mutual confirmation of the two data sets, see again table 3 and fig. 13.

Glass	Exponential fit for resonator measured tan(δ)	RMS relative deviation of resonator-measured tan(δ) from resonator based fit	RMS relative deviation of waveguide-determined tan(δ) from resonator based fit
AF32®eco	0.0048*Exp(0.0072*v/GHz)	25.4%	54.5%
Borofloat33®	0.0075*Exp(0.0046*v/GHz)	9.3%	53.1%
MEMpax®	0.009*Exp(0.0053*v/GHz)	-	-

Table 3: Exponential fits and RMS relative deviations of the loss angle in the frequency range 4GHz-77GHz for AF32®, Borofloat33®, and MEMpax® determined by resonator and waveguide methods. As above, one data pair from the 150GHz-210GHz measurement has been added for the exponential fit (180GHz, mean value).

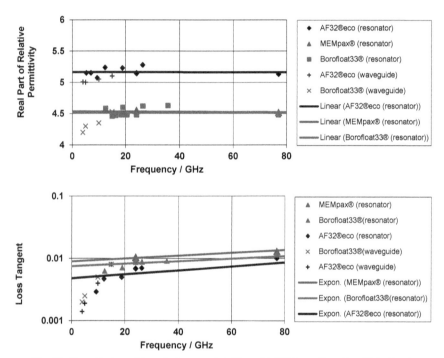

Fig. 13: Compilation of the relative permittivities of AF32®eco, Borofloat33®, □and MEMpax®□ obtained both from resonator and waveguide methods,□ in the 4GHz–77GHz range. As above, one data pair from the 150GHz-210GHz measurement has been added for the exponential fit.

CONCLUSION

Precise and consistent values for the complex permittivity of glass in the 1GHz–100GHz range have been obtained using both non metallized and metallized samples. This is an important result, since the former yield values characterizing the glass as it is provided by the manufacturer, while the latter exhibit those values "seen" by the designer.

Particularly interesting substrates are glasses which are commercially available as wafers and whose thermal expansion coefficients match the one of silicon, e.g. the alumosilicate glass AF32®eco as well as the borosilicate glasses Borofloat33® and MEMpax® from SCHOTT.

The values may be used, for instance, by radio frequency antenna designers who want to make use of the favourable properties these glasses provide, i.e., first, a small real part of the relative permittivity as compared to silicon, and, second, dielectric losses comparing to high resistivity silicon rather than to low resistivity silicon.

The reproducibility of the resonator-based measurements (split cylinder, open resonator; non-metallized samples) of the real part of the relative permittivity, a crucial value for system design, amounts to about 1%. As this reproducibility is affected by the dimensional stability of the samples the low value of about 1% gives evidence, first, of high precision measurement equipment in all cases and, second, of a very low "Total Thickness Variation" (TTV) for the single wafer as well as a very low wafer-to-wafer thickness variation.

The exact value of the loss tangent, on the contrary, is normally a less delicate quantity so that it often suffices to know that it does not exceed a certain threshold. For such a purpose, the (remarkable) consistency of 10-25% achieved by the resonator-based methods is sufficient.

The results have been confirmed by an indirect determination of the permittivity from comparing theoretical simulation and experimental characterization of special radio frequency waveguide devices on glass substrates (metallized samples).

The consistency with the resonator-based methods is 3-5% for the real part of the relative permittivity and ca. 50% for the loss tangent. Also the latter value is considered satisfactory with respect to the very small amount of these losses and the fact that determining them by the indirect method implies the calculation of small differences of big quantities (measured loss minus losses from other loss mechanisms).

REFERENCES

[1] A. Polyakov, P.M. Mendes, S.M. Sinaga, M. Bartek, B. Rejaei, J.H. Correia, and J.N. Burghartz, Processability and Electrical Characteristics of Glass Substrates for RF Wafer-Level Chip-Scale Packages, Proceedings of the IEEE Electronics and Technology Conference, **53**, 875–880 (2003).

[2] H. Zhang, M. Li, D. Zhang, N.C. Tien: A process research for integrated RF Tunable Filter, Proceedings of the IEEE International Conference on Nano/Micro Engineered and Molecular Systems, **1**, 659-661 (2006),

[3] F. Ohnimus, U. Maaß, G. Fotheringham, B. Curran, I. Ndip, T. Fritzsch, J.Wolf, S. Guttowski, K.-D. Lang: Design and Comparison of 24GHz Patch Antennas on Glass Substrates for Compact Wireless Sensor Nodes, International Journal of Microwave Science and Technology Volume 2010, Article ID 535307 (2010)

[4] M. Töpper, I. Ndip, R. Erxleben, L. Brusberg, N. Nissen, H. Schröder, H. Yamamoto, G. Todt, and H. Reichl: 3-D Thin Film Interposer Based on TGV (Through Glass Vias): An Alternative to Si-Interposer, Proceedings of the IEEE Electronic Components and Technology Conference, **60**, 66-73 (2010).

[5] V. Sukumaran, T. Bandyopadhyay, Q. Chen, N. Kumbhat, F. Liu, R. Pucha, Y. Sato, M. Watanabe, K. Kitaoka, M. Ono, Y. Suzuki, C. Karoui, C. Nopper, M. Swaminathan, V. Sundaram, R. Tummala: Design, fabrication and characterization of low-cost glass interposers with fine-pitch through-packages. Proceedings of the IEEE Electronic Components and Technology Conference, **61**, 583-588 (2011)

[6] U. Arz, J. Leinhos, and M.D. Janezic, "Broadband Dielectric Material Characterization: A Comparison of On-Wafer and Split-Cylinder Resonator Measurements", Proceedings of the European Microwave Conference, **38**, 571-574 (2008)

[7]T. Zwick, A. Chandrasekhar, C. W. Baks, U. R. Pfeiffer, S. Brebels, and B. P. Gaucher, Determination of the Complex Permittivity of Packaging Materials at Millimeter-Wave Frequencies, IEEE Transactions on Microwave Theory and Techniques, **54**, No. 3, 1001-1010 (2006)

[8]V. V. Parshin, R. Heidinger, and B. A. Andr, Silicon as an advanced window material for high power gyrotrons, *Int. J. Infrared Millim. Waves*, **16**, No. 5, 863–877 (1995)

[9]I. Ndip, B. Curran, S. Guttowski, and H. Reichl, Modeling and Quantification of Conventional and Coax-TSVs for RF Applications, Proceedings of European Microelectronics and Packaging Conference **2009**, 624-627, (2009)

[10]L.L.W.Leung, J. Zhang, W.C.Hon, K.J.Chen: High-Performance CMOS-Compatible Micromachined Edge-Suspended Coplanar Waveguides on Low-Resistivity Silicon Substrates, Proceedings of the European Gallium Arsenide Applications Symposium, **12**, 443-446 (2004).

[11]ASTM F723-99 Standard Practice for Conversion Between Resistivity and Dopant Density for Boron-Doped, Phosphorus-Doped, and Arsenic-Doped Silicon (Withdrawn 2003), ASTM International, West Conshohocken, PA, USA

[12]A. Polyakov, S.M. Sinaga, P.M. Mendes, M. Bartek, J.H. Correia, and J.N. Burghartz, High-Resistivity Polycrystalline Silicon as RF Substrate in Wafer-Level Packaging, *Electronic Letters* **41**, No. 2, 100-101 (2005)

[13]M.D. Janeciz, "Nondestructive Relative Permittivity and Loss Tangent Measurements Using a Split-Cylinder Resonator", PhD thesis, University of Colorado at Boulder, USA (2003).

[14]G. Fotheringham, C. Tschoban, F. Ohnimus, I. Ndip, S. Guttowski, K.-D. Lang, U. Fotheringham, M. Letz: Extraction of Dielectric Properties of Glass Substrates for RF Applications up to 40 GHz Using a Split-Cylinder-Resonator, Proceedings of the Smart Systems Integration Conference, **5**, (2011)

[15]A. Kilian: Funktionsintegration konformer Automobilradarantennen in mehrlagige Kunststoff-karosserieteile. Dissertation, University of Erlangen-Nürnberg, Germany (2011). Published by Verlag Dr. Hut, München (2012)

[16]F. Gumbmann, J. Weinzierl, H.P. Tran and L.-P. Schmidt, 3D Millimeterwellen-Abbildung von dielektrischen Probekörpern und numerische Rekonstruktion der Materialeigenschaften, Proceedings of the Annual Meeting of Deutsche Gesellschaft für Zerstörungsfreie Prüfung (2007)

[17] R.N. Simons, "Coplanar Waveguide Circuits, Components and Systems", Wiley Interscience, New York (2001)

FORMATION OF SILVER NANO PARTICLES IN PERCOLATIVE Ag-PbTiO$_3$ COMPOSITE DIELECTRIC THIN FILM

Tao Hu, Zongrong Wang, Liwen Tang, Ning Ma, Piyi Du*
State Key Laboratory of Silicon Materials, Department of Materials Science and Engineering,
Zhejiang University, Hangzhou, China, 310027

ABSTRACT

The silver nanoparticles dispersed PbTiO$_3$ thin film is synthesized by sol gel method. The influence of heat treatment conditions on the formation of silver was investigated in details. Results show that the size of silver particles is about 600 nm and 100 nm respectively in the rapid and slow heating process. With increasing heat treatment temperature, the number of silver particles increases and the size decreases during heat treated in air atmosphere. While the number and size both decease when thin films heat treated in H$_2$ atmosphere. And the decrease speed of size of silver particles is much faster for thin films heat treated in H$_2$ atmosphere than that for thin films heat treated in air atmosphere. The size of silver particles is about 12 nm and 5 nm for thin films heat treated in H$_2$ atmosphere at temperature of 550°C and 600°C respectively. The permittivity of thin film with silver nanoparticles is much higher than that without silver nanoparticles based on the percolation effect occurring in the silver nano particle dispersed Ag-PbTiO$_3$ thin film.

INTRODUCTION

With the miniaturization of electronic devices, more and more attention has been paid to the thin film materials with high permittivity. Many researches on dielectric matrix with conductive particles dispersed are reported to increase the permittivity of the matrix due to percolation effect.[1-7] The permittivity of composites is even several orders higher than that of pure dielectric materials.[8-10] In recent years, the preparation and property of metal dispersed dielectric composite thin films have attracted many interests. However, the realization of percolation effect in thin film matrix is not easy for its thickness is just hundreds of nanometers. Thus it demands that size of metal particles should be much smaller than the thickness of film, such as several nanometers. Or large sized conductive particles are capable of forming a conducting path across the upper and bottom electrodes normally used in the thin film devices and the percolation effect disappear. Meanwhile, the formation of metal particles is influenced by the preparation process such as heat treatment conditions for easy aggregation of metal particles with small size. Thus investigation of formation of metal particles in thin film matrix is very important for realization of percolation effect to improve the dielectric property of thin film materials.

In this paper, we proposed Ag as metal particles and PbTiO$_3$ thin film as matrix. And Ag-PbTiO$_3$ thin films were prepared by sol-gel method. The influence of heat treatment condition on formation of silver particles was investigated in details.

EXPERIMENTAL

The Ag-PbTiO$_3$ thin films were prepared in situ by sol-gel method with tetrabutyl titanate, lead nitrate and silver nitrate as raw material and with deionized water and concentrated nitric acid as catalyst and ethylene glycol monomethyl ether as solvent. Initially, Ti(C$_4$H$_9$O)$_4$ was dissolved in ethylene glycol monomethyl ether to form Ti solution. Pb(NO$_3$)$_2$ was dissolved in distilled water and ethylene glycol monomethyl ether (1:9 in volume ratio) to prepare Pb solution. Then PbTiO$_3$ sol precursor was obtained by mixing the two solutions. Finally, the silver nitrate powder and the precursor solution of PbTiO$_3$ were mixed with molar ratio of Ag/Ti (defined as x) of 0.5 to obtain Ag-Pb-Ti sol precursor for Ag-PbTiO$_3$ thin film. Initially, the wet layer was prepared using the Ag-Pb-Ti sol precursor by dip coating method with the withdraw speed of 4 cm/min on cleaned indium tin oxide/glass substrates. Then the wet layer was heat treated for 10 minutes to be thin film. The coating and heat treatment process was repeated to obtain Ag-PbTiO$_3$ thin films for controlling their thickness. The heat treatment temperature was controlled at 600 °C with rising speed of 1000 °C/min (rapid heating) and 10 °C /min (slow heating) from room temperature, respectively. Meanwhile, the thin films were annealed with rapid heating speed in air atmosphere and H$_2$ atmosphere (mixed of 97.5 vol% nitrogen and 2.5 vol% hydrogen) in a conventional muffle furnace with the mixed gas flux of 400 sccm at 250 °C, 350 °C, 450 °C, 550 °C and 600 °C, respectively.

The crystalline phase structure of the thin films was characterized by X-ray diffractometer (Rigaku D/max-rA) with Cu Kα radiation of 0.15418 nm. The morphology of the samples was observed in secondary electron imaging mode by scanning electron microscopy (SEM, HitachiSEM-S-570). The size of Ag nanoparticles in the thin films was measured by absorption spectra of ultraviolet – visible (UV – VIS) light using a spectrophotometer (Perkin – Elmer-Lamba-20). The dielectric constant was measured by LCZ meter (Kethley3330).

RESULTS

Fig. 1 shows XRD patterns of silver dispersed PbTiO$_3$ thin films, with molar ratio of Ag/Ti of 0.5, processed with slow and rapid heat treatment speed from room temperature to 600°C in air atmosphere. A set of the diffraction peaks of the perovskite phase of PbTiO$_3$ can be found in the samples with both the slow and rapid process. Other than the perovskite phase, there is a diffraction peak exhibited as the pyrochlore phase at about 29° in 2θ in sample prepared with the slow heating process while it is very weak in the sample with the rapid heat treatment. Moreover, a diffraction peak of cubic phase of silver appears at about 38° for the rapid heating sample and disappears in the slow heating one.

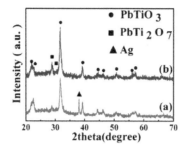

Fig. 1 XRD pattern of Ag dispersed PbTiO₃ with molar ratio of Ag/Ti of 0.5 heat treated in air atmosphere at 600 °C with different heating rate of (a) 1000 °C/min, (b)10 °C/min, respectively

Fig. 2 shows SEM images of thin films with different ascending rates of heat treatment temperature. It can be clearly seen that the two thin films are both composed of island particles (bright cluster) and base matrix (dark area). Actually, the bright particles are one of the constituent phases of silver and the dark matrix is PbTiO₃.[10] The average size of them is about 600 nm and 100 nm in the thin films with the rapid and slow heating process respectively. The former is much larger than latter.

Fig. 2 SEM images of Ag dispersed PbTiO₃ with molar ratio of Ag/Ti of 0.5 heat treated in air atmosphere at 600°C with different heating rate of (a)1000°C/min and (b) 10°C/min

Fig 3 shows SEM images of thin films with molar ratio of Ag/Ti of 0.5 heat treated in air atmosphere at different temperatures. The silver particles (bright cluster) can be found in the thin film when the heat treatment temperature is higher than 250 °C. The number of the particles increases and the average size decreases with increasing the heat treatment temperature.

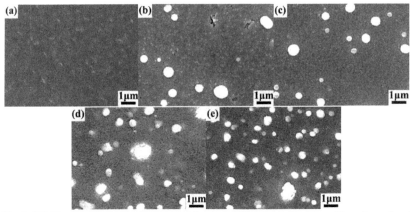

Fig. 3 SEM images of samples heat treated in air at different temperature with molar ratio of Ag/Ti of 0.5 (a-250 °C, b-350 °C, c-450 °C, d-550 °C, e-600 °C)

Fig. 4 shows SEM images of thin films with molar ratio of Ag/Ti of 0.5 heat treated in H$_2$ atmosphere at different temperatures. Large sized silver particles (bright cluster) can be found in the thin film when the heat treatment temperature is 250 °C and 350 °C and the number of silver particles increases with increase of heat treatment temperature from 250 °C and 350 °C. But, there are several particles with size of about 100 nm appearing in the thin films with heat treatment temperature of 450 °C and 550 °C. When the heat treatment temperature is 600 °C, the silver particles can not be observed.

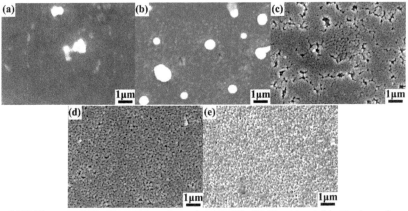

Fig.4 SEM images of thin films with molar ratio of Ag/Ti of 0.5 heat treated in H$_2$ atmosphere at different temperature. (a-250 °C, b-350 °C, c-450 °C, d-550 °C, e-600 °C)

Table I. Composition of Ag-dispersed lead titanate with molar ratio of Ag/Ti of 0.5 heat treated in air atmosphere at 600 °C with different ascending rate of heat treatment

Heat treatment process	The proportion of Ag:Pb:Ti
rapid heating	0.50:1.06:1
slow heating	0.42:0.95:1

DISCUSSION

As shown in XRD patterns of the thin films in Fig.1, apart from the perovskite phase of PbTiO$_3$, pyrochlore phase of lead titanate also forms in the composite thin films. The perovskite phase of PbTiO$_3$ seems to be formed perfectly with only a very little pyrochlore phase exhibited in the composite thin film with the rapid heating process. While the pyrochlore phase is much more in the thin film with slow heating process. According to the morphologies of the composite thin films as shown in SEM images in Fig.2, the Ag clusters are embedded typically in both the two kinds of thin films but the average size of them is about 600 nm and 100 nm respectively in the rapid and slow thermal processed thin films.

As is known during the sol gel preparation of the thin film, the decomposition and combustion of organic matter will initially occur and then crystallization begins with increasing heat treatment temperature. In this work, large amount of organic solvents was used during preperation of Ag-PbTiO$_3$ thin films. The organic solvents will consume oxygen in the furnace and decrease the partial pressure of oxygen around the thin films thus a local reducing atmosphere form during heat treatment process.[2,11] Ag$^+$ and Pb^{2+} will be reduced gradually to be Ag^0 and Pb^0 respectively under the local reducing environment formed by decomposition of organic matter. It is known, Pb^0 volatilizes easily with increasing temperature. Once the temperature reaches to about 600°C, the perovskite phase of PbTiO$_3$ would be formed easily and significantly. Pb^{2+} will participate directly and mainly to form lead titanate but not volatilize. For the thin film heated by the rapid heating process, since it maintains a very short time before reaching to high heat treatment temperature of 600°C, less Pb^{2+} will be reduced to be Pb^0 and the amount of lead volatilized is thus relatively small and the perovskite phase forms dominantly as shown in Fig.1(a). However, the high speed of increasing temperature from low to high of 600°C at which lead titanate forms speedily will let the lead titanate form sharply in the thin film. And the density of the speedily formed thin film would be probably low. The Ag^0 is thus easy to diffuse and move to form silver particle clusters with the high kinetic energy and migration ability of Ag^0 in the thin film. The average size of the silver clusters is larger up to about 600 nm as shown in Fig.2(a). Furthermore, when thin film is heated with slow heating process, it maintains a long time before getting to 600 °C. More Pb^{2+} may have therefore enough time to reduce to be Pb^0 and volatize. The pyrochlore phase forms easily in the thin film as shown in Fig. 1(b) due to less content of lead. It can be seen in Table. I in which EDX results are exhibited for the thin films without and with rapid thermal process. It shows that the composition molar ratio of the lead content in the thin film with slow thermal process is obviously lower than that in RTP thin film. But in RTP thin film, the fraction of lead is basically the same as that in the sol precursor. However in the slowly heated thin film, due to a long time modification and growth during increasing temperature, perfect crystalline phases and dense microstructure would appear probably at high temperature although the pyrochlore phase forms easily in the thin film. The

Ag^0 diffuses relatively small and thus the silver clusters forms not so easily. The size of the silver clusters is hence only about 100 nm which is much smaller than that in the thin film with rapid heating process.

Apparently, the size of the silver particles in the composite thin film can be controlled through adjusting the diffusion ability of Ag^0 and the nature of the thin film microstructure. The more perfect the crystalline phase in thin film and the lower the diffusion ability of Ag^0 are, the smaller the size of the silver particles embedded is. In the thin film with RTP in air atmosphere, Ag^0 diffuses and aggregates easily to form large sized silver clusters. Nevertheless in the thin film with slow thermal process, the diffusion of Ag^0 is not so easy and thus small sized clusters form. But in this case the pyrochlore phase appears significantly in the composite thin film which is actually not benefit to dielectric properties that is important for use as high performance dielectrics.

Moreover, the formation of the silver particles depends also on atmosphere during heat treatment. As shown in Fig.3, for thin films heat treated in air atmosphere, the number of silver particles increases and the average size decreases to about 600 nm with increasing heat treatment temperature to 600 °C. However, the number and the average size of silver particles both decrease with increasing heat treatment temperature for thin films heat treated in H_2 atmosphere and it finally disappears at 600 °C as shown in Fig.4.

Considering the preparation process of Ag-PbTiO₃ thin films in this work, the Ag^+ will be oxidized into Ag_2O first and then reduced into Ag^0 during the process of annealing.[12] The formation of silver particles is related to the Gibbs free energy of the reaction of decomposition of Ag_2O which can be described below:

$$\frac{1}{2} Ag_2O \rightarrow Ag + \frac{1}{4}O_2 \uparrow.$$

The Gibbs free energy (ΔG) is:

$$\Delta G = \Delta G_0 + \frac{1}{4} RT \ln P_{O2} - RT \ln(c_{Ag^+}) \tag{1}$$

In which ΔG$_0$ is the standard Gibbs free energy written as:[13]

$$\Delta G_0 = 16192 + 3.76T \ln(T) - 58.24T \tag{2}$$

where P_{O2} is the partial pressure of O_2, c_{Ag+} is concentration of Ag^+ which can be approximated as x/(1+x). The P_{O2} is the partial pressure of oxygen which is 0.2 and 10^{-6} for thin films heat treated in air and H_2 atmosphere respectively.

Fig. 5 shows Gibbs free energies of the decomposition reaction with molar ratio of Ag/Ti of 0.5 heat treated in air and H_2 atmosphere at different temperatures. For the thin film heat treated in air atmosphere with the heat treatment temperature of 250 °C, the Gibbs free energy is positive that there are no silver particles formed in the thin film. When the heat treatment temperature increases above 350 °C, the Gibbs free energy is negative and it decreases with increasing heat treatment temperature. The silver particles appear thus in the thin film due to occurrence of the reaction and the number of the particles increases with increasing the heat treatment temperature. It is because the Gibbs free energy is more negative with increasing the heat treated temperature and thus forming more easily. The content of the silver particles increase with increasing heat treatment temperature as shown in Fig. 3. However for thin film heat treated in H$_2$ atmosphere, the Gibbs free energy is negative at all temperatures as shown in Fig. 5, which is smaller than that in air atmosphere at the same temperature and it decreases also with increasing the heat treatment temperature. So the silver particles would appear in the thin films heat treated at all temperatures and the number of silver particles would increase with increasing heat treatment temperature, although the silver particles with the size of micrometer scale disappeared in the thin films heat treated in H$_2$ atmosphere at heat treatment temperature above 350 °C as shown in Fig. 4.

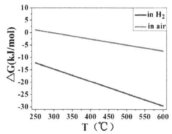

Fig. 5 Gibbs free energy for the decomposition reaction of Ag$_2$O to be Ag^0 in the thin films with molar ratio of Ag/Ti of 0.5 heat treated in air and H$_2$ atmosphere at different temperature

Actually in the case of thin films with molar ratio of Ag/Ti of 0.5 heat treated at different temperature in H$_2$ atmosphere, the silver particles may exist in the thin films with the dimension being down to nano scale. Hence, they can not be observed directly in the thin film using SEM measurement. It can be confirmed by UV-vis spectra as shown in Fig. 6. Compared with the thin films heat treated in air atmosphere, UV-vis spectra of shows obviously new absorption peak at about 410 nm for the thin films heat treated in H$_2$ atmosphere. It was recognized that the absorption peak at about 410 nm shown in UV-vis spectra is from surface plasma resonance of nano sized Ag particle and the area of the peak depends on the content of the Ag particles. It implies that there are nano silver particles formed in the thin films heat treated at temperature of 550 °C and 600 °C in H$_2$ atmosphere but it does not form in all the other thin films which are heat treated in air atmosphere[9,14,15] at different temperatures and heat treated in H$_2$ atmosphere below 550 °C. And the content of the nano sized Ag particles increases with increasing heat treatment temperature from 550 °C to 600 °C.

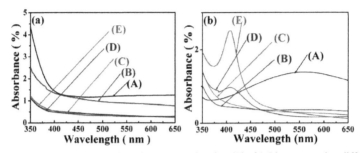

Fig. 6 UV-vis spectra of samples with molar ratio of Ag/Ti of 0.5 heat treated at different temperature in (a) air and (b) H$_2$ atmosphere. (A-250 °C, B-350 °C, C-450 °C, D-550 °C, E-600 °C)

Why does the nano silver particle form only in the thin films heat treated at temperature of 550 °C and 600 °C in H$_2$ atmosphere but not in others? It can be analyzed on the basis of nucleation and growth of silver particle.

Generally there are two ways for the nucleation of silver. One is that it nucleates Ag^0 directly[16,17] and the other is that it nucleates Ag$_3$$^{2+}$ after both Ag^0 and Ag$^+$ nearby form Ag$_3$$^{2+}$ in the thin film.[18] In general, Ag$_3$$^{2+}$ can be detected by the PL spectra because it arouse the recombination process of indirect transition between belt of X$_4$'->L$_2$', emitting a fluorescence of wavelength of 500-600 nm after irradiated by the UV light of 200-400 nm.[19,20] However for thin films heat treated in both air and H$_2$ atmosphere at different temperatures as shown in PL spectra in Fig. 7, there are no emission peaks appearing at wavelength of 500-600 nm. It is clear that Ag^0 is the dominant in nucleation of silver in this case. Using the traditional equations below:

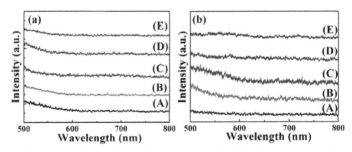

Fig. 7 shows PL spectra of thin films with molar ratio of Ag/Ti of 0.5 heat treated in air (a) and H$_2$ (b) atmosphere at different temperature. (A-250°C, B-350°C, C-450°C, D-550°C, E-600°C)

$$I = a\gamma_0 n \exp\left[-\left(\Delta G_{a,1} + \frac{16\pi\gamma_{LS}^3}{3(\Delta G_V)^2}\right)/RT\right] \qquad (3)$$

$$U = f\gamma_0 \exp(-\frac{\Delta G_{a,1}}{RT})\left[1-\exp(-\frac{\Delta G_V}{RT})\right] \tag{4}$$

where I is the nucleation speed, U is growth speed, a and f are constants, γ_0 is the transition frequency of silver, n is the atom concentration for nucleation, γ_{LS} is the interfacial energy of silver nucleation and the matrix of lead titanate, $\Delta G_{a,1}$ is activation energy for the diffusion of dissociative Ag^0 to the nucleation spot, and ΔGv is the free energy difference between the dissociative Ag^0 and the silver particle. Considering that only the part of Ag^0 having migrating ability can participate in nucleation process, we can calculate the n as shown below,

$$n = n_0 \exp(-\frac{\Delta G_r}{RT})\exp\left(-\frac{\Delta G_{a,2}}{RT}\right) \tag{5}$$

Where ΔG_r is the activation energy of the formation of Ag^0 from Ag_2O, $\Delta G_{a,2}$ is the activation energy for the diffusion of dissociative Ag^0 in the matrix and n_0 is the consistence of silver ion in the thin films.

Generally, more atoms participate in the nucleation process then fewer atoms will participate in the growth process when the concentration of silver atom is fixed, which leads to decreasing the size of the grains. Thus we can use the ratio of I over U (defined as B) to investigate the size of the silver particles. The value B can be expressed as follow:

$$B = \frac{I}{U} = \frac{an_0}{f}\exp\left[-\left(\Delta G_r + \Delta G_{a,2} + \frac{16\pi\gamma_{LS}^3}{3(\Delta G_V)^2}\right)/RT\right]/\left[1-\exp(-\frac{\Delta G_V}{RT})\right] \tag{6}$$

In which

$$F = \exp\left[-\left(\Delta G_r + \Delta G_{a,2} + \frac{16\pi\gamma_{LS}^3}{3(\Delta G_V)^2}\right)/RT\right]/\left[1-\exp(-\frac{\Delta G_V}{RT})\right] \tag{7}$$

It is difficult to calculate the value of B quantificationally, but we can calculate the value F to estimate the change of B.

The parameter of $\Delta G_{a,2}$, γ_{LS} and ΔG_v for the diffusion of silver in the matrix of lead titanate can be substituted approximately by that for the diffusion of silver in the matrix of Pb(Ni₁/₃Nb₂/₃)O₃-PbTiO₃-PbZrO₃ for their structure and properties are close, thus $\Delta G_{a,2}$= 168 kJ/mol, γ_{LS}= 0.0005T+0.4653 J/m2, ΔGv= 90600-0.9RT J/mol.[21] In addition, ΔG_r for the thin films heat treated in air and H_2 atmosphere is 39 kcal/mol and 15 kcal/mol, respectively.[22,23]

Considering the nucleation and growth processes are thermally active, the nucleation rate should be also thermally active and then B and F can be expressed to be as following:

$$B = B_0 \exp(-\frac{Q}{RT}) \tag{8}$$

$$F = \frac{an_0}{f} B_0 \exp(-\frac{Q}{RT}) = F_0 \exp(-\frac{Q}{RT}) \tag{9}$$

where B_0 and F_0 are pre-exponential factors and Q is the active energy.

Fig. 8 shows relationship of lnF and 1/T for thin films with molar ratio of Ag/Ti of 0.5 heat treated at different temperature in air and H_2 atmospheres respectively. It indicates that the value of F for thin films heat treated in H_2 atmosphere is bigger than that heat treated in air atmosphere at all temperatures. And the relationship between lnF and 1/T are both linear, which means that the nucleation rate is thermally active. In addition, the active energy is 3.34 kJ/mol and 4.79 kJ/mol for thin films heat treated in H_2 and air atmosphere respectively as shown in Fig.8, which is smaller for thin films heat treated in H_2 than that in air.

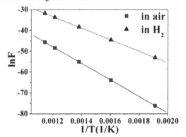

Fig. 8 Relationship of lnF and 1/T for thin films with molar ratio of Ag/Ti of 0.5 heat treated at different temperature in air atmosphere.

Generally, high ratio of I over U will result in small size of particles during formation process. In this case, ratio of I over U increases with increasing heat treatment temperature for thin films heat treated in both the atmospheres. Thus the average size of silver particles decreases in thin films heat treated in both the atmospheres with increasing heat treatment temperature as shown in Fig. 3 and Fig.4. In addition, it is bigger for thin films heat treated in H_2 atmosphere than that heat treated in air atmosphere at all temperatures. So the average size of silver particles is smaller in thin films heat treat in H_2 than that in thin films heat treated in air at all temperatures as shown also in Fig. 3 and Fig. 4. Moreover, the decrease speed of the average size of the silver particles is different in thin films heat treated in H_2 and air atmosphere and it is bigger in H_2 than that in air.

Fig. 9 shows relationship of lnF and lnS (S means average size of silver particles) for thin films with molar ratio of Ag/Ti of 0.5 heat treated in air and H_2 atmosphere. It indicates that relationship between lnF and lnS is linear and the slope is -3.6 and -42.5 for thin films heat treated in H_2 and air respectively. It is obvious that the size changes so large as the F changes only a little in the thin film heated in H_2. So the influence of variation of F on the size of silver particles is much stronger in the thin film heated in H_2 than that in air atmosphere. Therefore the size of silver particles decreases significantly to be about 12 nm and 5 nm which calculated from the absorption peak[24] in Fig. 7b at temperature of 550 °C and 600 °C respectively. It can be summarized that the value F that is the ratio of I over U can be used to characterize the size of silver particles commendably.

Considering that the only difference between the two atmosphere is $\triangle G_r$ on basis of Eq.(7), we can state that the formation of silver nanoparticles is controlled by the activation energy of formation of Ag^0 from Ag_2O. The smaller the active energy of formation of Ag^0 from Ag_2O, the smaller the size of silver particles will be. Actually the activation energy is much smaller for the thin film sintered in H_2 than that in air. That is why the nano sized Ag particles form easily in the thin film heated in H_2 but not in air atmosphere.

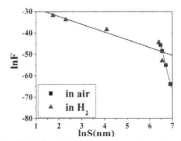

Fig.9 Relationship of lnF and lnS for thin films with molar ratio of Ag/Ti of 0.5 heat treated at different temperature in air atmosphere.

Fig. 10 shows the permittivity of the Ag- PbTiO₃ thin films heat treated in H_2 atmosphere at temperature from 450 °C to 600 °C and single phased PbTiO₃ thin film heat treated at 600 °C. It indicates that the permittivity of thin film with silver nanoparticles is much higher than that of thin film without silver nano particles. Actually the increase in permittivity of thin film with silver nanoparticles dispersed is due to the percolation behavior appearing in the Ag- PbTiO₃ composite.[6] The percolation effect can be attributed to the existence of metal particles inside the thin film and insulation of the conductive particles by barriers of dielectric material. Many micro capacitors with very short distance between electrodes (i.e. silver particles) thus form. The more the micro capacitors and the shorter the distance between particles are, the higher the permittivity is. The size of silver particles decreases thus amount of micro capacitors increases with increasing heating temperature from 450 to 600 °C. The permittivity thus increases in the case for increasing amount of micro capacitors with short distance between electrods with increasing heating temperature from 450 to 600 °C. And the permittivity of thin films heat treated at temperature of 550 and 600 °C is much higher than that of pure PbTiO₃ thin film for exsistance of silver nanoparticles dispersed. While the permittivity of Ag- PbTiO₃ thin film heat treated at temperature of 450 °C is a little lower than that of pure PbTiO₃ thin film may be for not good formation of the matrix. Obviously, the percolation behavior contributes dominantly the high permittivity of the silver nanoparticles dispersed thin film.

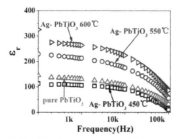

Fig. 10 permittivity of Ag- PbTiO₃ thin films heat treated in H₂ atmosphere at temperature from 450 °C to 600 °C and pure PbTiO₃ thin film heat treated at 600 °C.

CONCLUSIONS

The silver nanoparticles dispersed PbTiO₃ thin film is synthesized by sol gel method when it is heat treated in H_2 atmosphere with heat treatment temperature higher than 550 °C. The formation of silver particles is influenced by the heat treatment conditions. The size of silver particles was 100 nm in the rapid heating process which is smaller than that in slow heating process for harder diffusion of silver. With increasing heat treatment temperature, the silver particles formed easily thus the content of silver particles increases. The size of silver particles decreases with increasing heat treatment temperature in air and H_2 atmosphere for the increase of ratio of the nucleation speed over the growth speed. Moreover the size of silver particles deceases more speedily when thin films heat treated in H_2 and it decreases to be about 12 nm and 5 nm at temperature of 550 °C and 600 °C. Actually the formation of silver nano particles is controlled by the active energy of formation of Ag^0 from Ag_2O. The smaller the active energy of formation of Ag^0 from Ag_2O is, the smaller the size of silver particles will be. The permittivity of thin film with silver nanoparticles is much higher than that of thin film without silver nano particles due to the percolation effect.

ACKNOWLEDGEMENT

This work is supported by NSFC(51172202, 50872120), ZJ-NSF(Z4110040), the National Key Scientific and Technological Project (2009CB623302), and Fundamental Research Funds for the Central Universities, China, respectively.

REFERENCES

[1]M.S. Wang, J.L. Zhu, W.L. Zhu, B. Zhu, J. Liu, X.H. Zhu, Y.T. Pu, P. Sun, Z.F. Zeng, X.H. Li, D.Q. Yuan, S,Y. Zhu, and G. Pezzotti, The Formation of Percolative Composites with a High Dielectric Constant and High Conductivity, *Angew. Chem. Int. Ed.,* **51**, 9123-7 (2012)

[2]T, Hu, Z.R. Wang, Y.B. Su, L.W. Tang, G. Shen, C.L. Song, G.R. Han, W.J.Weng, N. Ma, and P.Y. Du, Formation of Ag Nanoparticles in Percolative Ag- PbTiO₃ Composite Thin Films through Lead-rich Ag- Pb Alloy Particles Formed as Transitional Phase, *Thin Solid Films,* **524**, 121-6 (2012).

[3]S. Roy and S.B. Majumder, Percolative Dielectric Behavior of Wet Chemical Synthesized Lead Lanthanum Titanate-cobalt Iron Oxide Composite Thin Films, *Phys. Lett. A,* **375**, 1538-42

(2011).

[4]J.Q. Huang, P.Y. Du, L.X. Hong, Y.L. Dong, and M.C. Hong, A Novel Percolative Ferromagnetic–Ferroelectric Composite with Significant Dielectric and Magnetic Properties, *Adv. Mater.*, **19**, 437-+ (2007).

[5]J.K. Yuan, W.L. Li, S.H. Yao, Y.Q. Lin, A. Sylvestre, and J.B. Bai, High Dielectric Permittivity and Low Percolation Threshold in Polymer Composites Based on SiC-carbon Nanotubes Micro/nanoHybrid, *Appl. Phys. Lett.*, **98**, 032901 (2011).

[6]Z.R. Wang, T. Hu, X.G. Li, G.R. Han, W.J. Weng, N. Ma, and Piyi Du, Nano Conductive Particle Dispersed Percolative Thin Film Ceramics with High Permittivity and High Tunability, *Appl. Phys. Lett.*, **100**, 132909 (2012).

[7]M. Molberg, D. Crespy, P. Rupper, F. Nuesch, J.A.E. Manson, C. Lowe, and D.M. Opris, High Breakdown Field Dielectric Elastomer Actuators Using Encapsulated Polyaniline as High Dielectric Constant Filler, *Adv. Funct. Mater.*, **20**, 3280-3291 (2010).

[8]H. Zheng, Y.L. Dong, X. Wang, W.J. Weng, G.R. Han, N. Ma, and P.Y. Du, Super High Threshold Percolative Ferroelectric/ferrimagnetic Composite Ceramics with Outstanding Permittivity and Iitial Permeability, *Angew. Chem. Int. Ed.*, **48**, 8927-30 (2009).

[9]Z.R. Wang, T. Hu, L.W. Tang, N. Ma, C.L. Song, G.R. Han, W.J. Weng, and P.Y. Du, Ag Nanoparticle Dispersed PbTiO$_3$ Percolative Composite Thin Film with High Permittivity, *Appl. Phys. Lett.*, **93**, 222901 (2008).

[10]Z.R. Wang, T.Hu, L.W. Tang, C.L. Song, G.R. Han, W.J. Weng, N. Ma and P.Y. Du, Effect of Heat Treatment Temperature on the Formation of Ag Nanoparticles in Ag-PbTiO$_3$ Composite Thin Films, *Ferroelectrics*, **387**, 161-6 (2009).

[11]M.L. Calzada, J. Mendiola, F. Carmona, and R. Sirera, Pyrochlore-perovskite Phase Development of Sol-gel Derived Modified Lead Titanate Thin Films, *Microelectron. Eng.*, **29**, 197-200 (1995).

[12]B. Breitscheidel, J. Zieder, and U. Schubert, Metal Complexes in Inorganic Matrices. 7. Nanometer-sized, Uniform Metal Particles in a SiO$_2$ Matrix by Sol-gel Processing of Metal Complexes, *Chem. Mater.*, **3**, 559-66 (1991).

[13]H. J. Bi, W. P. Cai, C. X. Kan, L. D. Zhang, and D. Martin, Optical Study of Redox Process of Ag Nanoparticles at High Temperature, *J. Appl. Phys.*, **92**, 7491-7 (2002)

[14]R.C. Jin, Y.W. Cao, C.A. Mirkin, K.L. Kelly, G.C. Schatz, and J.G. Zheng, Photoinduced Conversion of Silver Nanospheres to Nanoprisms, *Science*, **294**, 1901-3 (2001).

[15]Q.F. Wang, H.J. Yu, L. Zhong, J.Q. Liu, J.Q. Sun, and J.C. Shen, Incorporation of Silver Ions into Ultrathin Titanium Phosphate Films: In Situ Reduction to Prepare Silver Nanoparticles and Their Antibacterial Activity, *Chem. Mater.*, **18**, 1988-94 (2006).

[16]S.E. Paje, J.L. Lopis, M.A. Villegas, M.A. Garcyal, and J.M. Fernandez Navarro, Thermal Effects on Optical Properties of Silver Ruby Glass, *Appl. Phys. A*, **67**, 429-33 (1998).

[17]M.A. Garcia, S.E. Paje, J.L. Lopis, M.A. Villegas, M.A. Garcyal, and J.M. Fernandez Navarro, Optical Spectroscopy of Arsenic and Silver ContainingSol-gel Coatings, *J Phys D: Appl. Phys.*, **32**, 975-80 (1999).

[18]E. Borsella, E. Cattaruzza, G. De Marchi, F. Gonella, G. Mattei, P. Mazzoldi, A. Quaranta, G. Battaglin, and R. Polloni, Synthesis of Silver Clusters in Silica-based Glasses for Optoelectronics Applications, *J. Non-Crys. Solids*, **245**, 122-8 (1999).

[19]M.G. Garnica-romo, J.M. Limon Yanez, and J. Gonzalez-Hernandea, Sturcture and Electorn Spin Resonance of Annealed Sol-gel Containing Ag, *J. Sol-Gel Sci. Technol.*, **24**, 105-12 (2002).
[20]E. Borsella, G. Battaglin, M.A. Garcia, F. Gonella, P. Mazzoldi, R. Polloni, and A. Quaranta, Structural Incorporation of Silver in Soda-lime Glass by the Ion-exchange Process: a Photoluminescence Spectroscopy Study, *Appl. Phys. A: Materials Science & Processing*, **71**, 125-32 (2000)..
[21]J.L. Daniel, G. Devendra, R.N. Michael, and I. Yoshihiko, Diffusion of 110mAg Tracer in Polycrystalline and Single-Crystal Lead-Containing Piezoelectric Ceramics, *J. Am. Ceram. Soc.*, **84**, 1777-84 (2001).
[22]A. Van Tiggelen, L. Vanreusel, P. Neven, Contribution à l'étude des Réactions entre Gaz et Solides I.-Réactions entre les sels d'argent et l'hydrogène, *Bull. Soc. Chim. Belg.*, **61**, 651-82 (1952).
[23]J. Halpen, G. Czapski, J. Jortner, and G. Stein, Mechanism of the Oxidation andReduction of Metal Ions by Hydrogen Atoms, *Nature*, **186**, 629-30 (1960).
[24]G.W. Arnold, Near-surface Nucleation and Crystallization of an Ionimplanted Lithia-alumina-silica Glass, *J. Appl. Phys.*, **46**, 4466-73 (1975).

SOFTWARE FOR CALCULATING PERMITTIVITY OF RESONATORS: HakCol & ErCalc

Rick Ubic
Boise State University
Boise, ID, USA

ABSTRACT
        The HakCol and ErCalc programmes are designed to quickly and accurately calculate the relative permittivity of microwave dielectric resonators. The HakCol programme assumes a simple Hakki-Coleman style parallel-plate dielectric-post geometry resonating in the TE_{011} mode, which is typically used to measure low-loss samples. The only inputs required are the cylindrical geometry of the resonator (diameter and thickness) and the resonant frequency. The ErCalc programme instead assumes a $TE_{01\delta}$ mode microwave shielded resonator in a cavity, in which case the required inputs also include the geometry of the test cavity and the dielectric constants of the surroundings. The latter technique is generally intended for very precise complex permittivity measurements of bulk low-loss cylindrical resonators. Some existing algorithms use a curve-fit to estimate the *ratio* of Bessel functions and modified Bessel functions in the isotropic resonator equation, leading to a small but measurable error. The chief advantage of the HakCol and ErCalc programmes is that all of the various Bessel functions are numerically calculated individually for an improved accuracy. The use of both programmes will be discussed, their algorithms analysed, and their outputs compared to alternative software currently available.

INTRODUCTION

Resonators in free space
        When microwaves enter a dielectric medium, they are slowed down by a factor equal to $\varepsilon_r^{-\frac{1}{2}}$; therefore:

$$\lambda_d = \frac{\lambda_0}{\sqrt{\varepsilon_r}} = \frac{c}{v\sqrt{\varepsilon_r}} \quad \therefore \quad v = \frac{c}{\lambda_d\sqrt{\varepsilon_r}} \tag{1}$$

At resonant frequency, $v = f_0$ and $\lambda_d \sim D$ (diameter of resonator); therefore:

$$f_0 = \frac{c}{D\sqrt{\varepsilon_r}} \quad \therefore \quad \varepsilon_r = \left(\frac{c}{Df_0}\right)^2 \tag{2}$$

Equation 2 is only valid in the case of resonators in free space. It fails for resonators in more realistic situations (e.g., on microstrips, in cavities, between shorting plates, etc.) In order to calculate permittivity in these geometries, several techniques have been developed and variously discussed. Perturbation techniques rely on the shift of f_0 (and Q) of a resonant cavity caused by the presence of a dielectric disk or sphere. Optical methods at microwave frequencies are suited to measurements at which $\lambda < 1$ cm and require a large amount of material. Transmission-line methods have the practical difficulty of requiring a very small waveguide for $\lambda < 4$ mm. All of these methods have an accuracy of approximately ±1%.

Parallel-plate dielectric-post method

The exact resonance method proposed by Karpova[1] and further developed by Hakki and Coleman,[2] Courtney,[3] and others yields errors of only ±0.1%. Karpova[1] used a re-entrant cavity (fig. 1) for the measurement of dielectric properties, but the physical size of the resonant structure required could be problematic for the low-mm range.

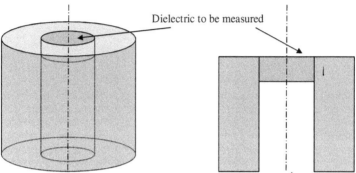

Dielectric to be measured

Figure 1. The re-entrant cavity reported by Karpova[1]

In order to avoid the problem of physical size whilst maintaining high accuracy, Hakki and Coleman[2] instead proposed an open-boundary resonant structure in which a dielectric rod was positioned between much larger conducting plates (fig. 2).

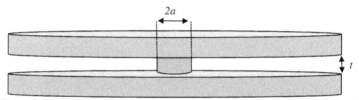

Figure 2. The open-boundary resonant structure proposed by Hakki and Coleman[2]

The characteristic equation which describes this condition for an isotropic resonator in a TE_{0mp} mode for which $\mu_r = \mu_0$ is:

$$\alpha \frac{J_0(\alpha)}{J_1(\alpha)} = -\beta \frac{K_0(\beta)}{K_1(\beta)}$$

(3)

where $J_0(\alpha)$ and $J_1(\alpha)$ are Bessel functions of the first kind of orders zero and one, respectively; and $K_0(\beta)$ and $K_1(\beta)$ are modified Bessel functions of the second kind of orders zero and one, respectively. The parameters α and β are functions of geometry, resonant wavelength, and permittivity:

$$\alpha = \frac{2\pi a}{\lambda_0} \sqrt{\varepsilon_r - \left(\frac{c}{v_p}\right)^2} \tag{4}$$

$$\beta = \frac{2\pi a}{\lambda_0} \sqrt{\left(\frac{c}{v_p}\right)^2 - 1} \tag{5}$$

where c is the speed of light, a is resonator radius as defined in Fig. 2, and v_p is the phase velocity in the resonator such that:

$$\left(\frac{c}{v_p}\right) = \left(\frac{p\lambda_0}{2t}\right) \tag{6}$$

where p is the number of longitudinal variations of field along the axis and $\lambda_0 = c / f_0$. Clearly v_p can be calculated from thickness and resonant frequency alone; and β can then be calculated from v_p, frequency, and radius.

The characteristic equation (equation 3) is transcendental, requiring a graphical solution. Hakki and Coleman[2] used analogue mode charts to relate various $\{\alpha_m\}$ to each corresponding value of β, resulting in somewhat limited accuracy.

Although this technique is sometimes called the Courtney method, "Courtney, actually, only perfected and scrutinized a parallel-plate arrangement introduced [10 years] earlier by Hakki and Coleman."[4] Courtney also adapted the technique to the use of coaxial probes (an innovation introduced four years earlier by Cohn and Kelly,[5] allowing a greater range of sample dimensions.

Shielded resonator in dielectric-rod waveguide

In the case of a shielded high-Q material in a resonant cavity, such as that shown in fig. 3, most of the electrical field is contained within the resonator itself (region 6), and very little exists in regions 1 and 2, and even less in regions 3 and 5. To a fair first approximation, then, the fields in regions 3 and 5 can be ignored.

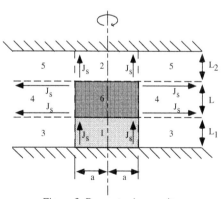

Figure 3. Resonator in a cavity

For the TE$_{01\delta}$ modes the requirement for continuity of fields leads to two simultaneous eigenvalue equations:

$$\frac{J_o(k_{\rho 1}a)}{J_1(k_{\rho 1}a)} = -\frac{k_{\rho 2}}{k_{\rho 1}} \cdot \frac{K_0(k_{\rho 2}a)}{K_1(k_2 a)}$$

(7)

$$\beta L = \frac{\phi_1}{2} + \frac{\phi_2}{2} + \ell\pi \qquad \ell = 0,1,2,3...$$

(8)

The radius of the resonator is a, as defined in fig. 3. The symbol k represents the radial propagation constants in the different regions of the model, which are functions of both frequency and dielectric constant; and ρ is the radial distance from the geometric centre. The arguments of the various Bessel functions are the eigenvalues of the system, where $k_{\rho 1}a$ is called the eigenvalue of the TE$_{0n}$ mode, and $k_{\rho 2}$ is given by:

$$k_{\rho 2} = \sqrt{k_0^{\;2}(\varepsilon_{r6} - \varepsilon_{r4}) - k_{\rho 1}^{\;2}}$$

(9)

where k_0 is called variously the *propagation constant*, *wavenumber*, or *phase constant* of free space, and has units of m^{-1}:

$$k_0 = \omega_0\sqrt{\varepsilon_0\mu_0}$$

(10)

In equation 8, β is the propagation constant of the resonator. If p is the number of axial variations of the field along the resonator's height, then $p = \ell + \delta$, where ℓ is an integer and δ is a non-integer number smaller than unity which depends in a complicated way on propagation constants and geometry. Whereas for the TE$_{011}$ mode discussed above, $\ell = 1$ and $\delta = 0$, for the TE$_{01\delta}$ mode, $\ell = 0$ and $\delta \neq 0$ The propagation constant common to both regions 4 and 6 is:

$$\beta = \sqrt{k_o^2\varepsilon_{r6} - k_{\rho 1}^2}$$

(11)

The symbols ϕ_1 and ϕ_2 are called the *phase angles* and are complex hyperbolic functions of the cavity geometry and the propagation constant of the resonator:

$$\phi_1 = 2\tan^{-1}\left(\frac{\alpha_1}{\beta}\coth\alpha_1 L_1\right)$$

(12)

$$\phi_2 = 2\tan^{-1}\left(\frac{\alpha_2}{\beta}\coth\alpha_2 L_2\right)$$

(13)

Here, α_1 and α_2 are called the *attenuation constants* above and below the resonator. These constants can be calculated directly from the frequency and radius, thus:

$$\alpha_1 = \sqrt{k_{\rho 1}^2 - k_o^2\varepsilon_{r1}}$$

(14)

$$\alpha_2 = \sqrt{k_{\rho 1}^2 - k_o^2\varepsilon_{r2}}$$

(15)

In these equations, ε_{r1} and ε_{r2} are the dielectric constants of the material below (the support) and above (air) the resonator.

It is interesting to note that equation 7 is similar to that developed by Hakki and Coleman[2] for their test geometry.

CALCULATIONS

Parallel-plate dielectric-post method (HakCol)

An improvement in accuracy over a purely graphical approach can be achieved by numerically solving for each Bessel/modified Bessel function rather than trying to read values off the mode charts of Hakki and Coleman[2] or even relying on curve fits. Ordinary Bessel functions of the first kind (fig. 4) can be numerically calculated according to:

$$J_v(z) = \sum_{r=0}^{\infty} \frac{(-1)^r \left(\frac{z}{2}\right)^{v+2r}}{r!\,\Gamma(v+r+1)} \tag{16}$$

in which the gamma (Γ) function is defined by:

$$\Gamma(x) = \int_0^{\infty} e^{-t} t^{x-1} dt \tag{17}$$

It is not possible to solve this integral explicitly, but a few ways exist to numerically calculate it. For this work, a simple rule for integral arguments ($x \geq 1$) suffices:

$$\Gamma(x) = (x-1)! \tag{18}$$

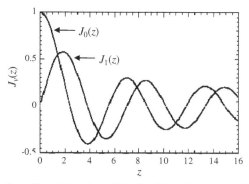

Figure 4. Ordinary Bessel functions of the first kind of orders 0 and 1 for $\alpha \leq 16$

Similarly, modified Bessel functions of the second kind (fig. 5) can be calculated according to the general equation:

$$K_v(z) = \frac{I_{-v}(z) - I_v(z)}{\left(\frac{2}{\pi}\right)\sin v\pi} \tag{19}$$

where $I_v(z)$ is the modified Bessel function of the first kind of order v:

$$I_v(z) = \sum_{r=0}^{\infty} \frac{1}{r!\Gamma(v+r+1)} \left(\frac{z}{2}\right)^{2r+v} \tag{20}$$

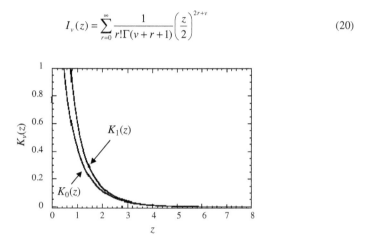

Figure 5. Modified Bessel functions of the second kind of orders 0 and 1 for $\beta \leq 8$

Using the general equations 16 and 19, the specific functions needed to determine the dielectric constant can be calculated as:

$$J_0(z) = \sum_{r=0}^{\infty} \frac{(-1)^r z^{2r}}{2^{2r}(r!)^2} \tag{20}$$

$$J_1(z) = \sum_{r=0}^{\infty} \frac{(-1)^r z^{2r+1}}{2^{2r+1} r!(r+1)!} \tag{21}$$

$$K_0(z) = -(\ln \tfrac{z}{2} + \gamma)I_0(z) + \sum_{r=1}^{\infty} \frac{\left(\tfrac{z}{2}\right)^{2r}}{r!r!} \left(\sum_{n=1}^{r} \frac{1}{n}\right) \text{ where } I_o(z) = \sum_{n=0}^{\infty} \frac{z^{2n}}{2^{2n}(n!)^2} \tag{22}$$

$$K_1(z) = (\ln \tfrac{z}{2} + \gamma)I_1(z) + \tfrac{1}{z} - \tfrac{1}{2}\sum_{r=0}^{\infty} \frac{\left(\tfrac{z}{2}\right)^{2r+1}}{r!(r+1)!} \left[2\left(\sum_{n=1}^{r} \frac{1}{n}\right) + \frac{1}{r+1}\right] \text{ where } I_1(z) = \sum_{n=0}^{\infty} \frac{z^{2n+1}}{2^{2n+1} n!(n+1)!} \tag{23}$$

with the first term in the series on the right-hand side of equation 23 being $z/2$. The γ needed in the modified Bessel function expressions (equations 22 and 23) is Euler's constant ($\gamma \sim 0.5772$). Obviously, the summations involved cannot possibly be evaluated to infinity; but for Bessel functions with small arguments (z) the value starts to converge after only ten iterations or so, and 20 iterations is sufficient for errors less than 1×10^{-6} up to $z = 16$. For the same kind of accuracy, 30 iterations are required for the summations involved in calculating the modified Bessel functions of orders zero ($z \leq 10$) and one ($z \leq 13$). The accuracy is best for low values of z.

  With these numerical methods it is possible now to solve equation 3 for $\beta \leq 10$. The algorithm employed in the HakCol programme[6] starts by calculating β from the resonator radius and resonant frequency. Next an approximate corresponding value for α is calculated using a curve fit ($R^2 = 0.9999929$) to the $m = 1$ mode chart of Hakki and Coleman,[2] as shown in fig. 6.

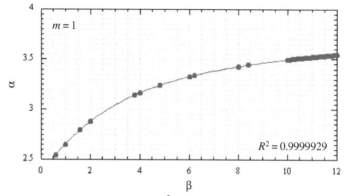

Figure 6. Curve fit to the mode chart[2] corresponding to $m = 1$ (TE_{01p} mode)

The polynomial which describes the curve in fig. 6 is:

$$\alpha = 2.3508 + 0.34969\beta - 0.051220\beta^2 + 0.0044392\beta^3 - 0.00020633\beta^4 + 3.9411 \times 10^{-6}\beta^5 \quad (24)$$

The α so calculated is used as a first approximation in order to calculate ε_r. Next, equation 3 is evaluated and if the two sides are unequal then ε_r is adjusted accordingly, α re-calculated, and the process iterates until equation 3 is satisfied.

In the event that $\beta > 10$, the numerical calculation of $K_0(\beta)$ fails due to overflow errors, but a different curve fit to the appropriate mode chart can be used to extend the range of the algorithm (Fig. 7). In the range $10 \le \beta \le 12$, a second-order polynomial curve fit to 21 evenly spaced data points provides sufficient accuracy ($R^2 = 0.99999734$) for reliable ε_r calculations:

$$\alpha = 3.0268228 + 0.065626204\beta - 0.0018728591\beta^2 \quad (25)$$

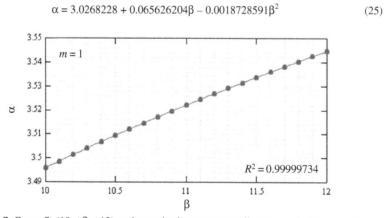

Figure 7. Curve fit ($10 \le \beta \le 12$) to the mode chart corresponding to $m = 1$ (TE_{01p} mode)

Shielded resonator in dielectric-rod waveguide (ErCalc)

For a high-Q material in a cavity, as proposed by Itoh and Rudokas[7] and modified by Kajez and Guillon[4] (fig. 3), most of the electrical field is contained within the resonator itself[7] (region 6), and very little exists in regions 1 and 2, and even less in regions 3 and 5. To a fair first approximation, then, the fields in regions 3 and 5 can be ignored. The solution to equations 7 and 8 was obtained using an algorithm based on that published by Kajfez and Guillon.[4] To start, the hyperbolic tangents in equations 12 and 13 can be calculated by definition:

$$\coth x = \frac{e^x + e^{-x}}{e^x - e^{-x}} \tag{26}$$

Next, equations 27 and 28 were used as generating functions:

$$f = \frac{J_0(x)}{J_1(x)} + \frac{z}{x} \frac{K_0(z)}{K_1(z)} \tag{27}$$

$$g = \beta L - \frac{\phi_1}{2} - \frac{\phi_2}{2} - \ell\pi \tag{28}$$

where $x = k_{\rho 1}a$, $y = k_0 a$, and $z = k_{\rho 2}a$. There are only two independent parameters involved, as z is linearly dependent on x and y. The parameter y is a function of frequency and radius only, neither of which change, so it is always known. The values of x and z must be guessed at first, $x = 3.2$, $z = 2.4$. In the vicinity of the solution, the functions f and g will be approximately linear:

$$f(x, y) \approx ax + by + c \tag{29}$$
$$g(x, y) \approx Ax + By + C \tag{30}$$

and setting $f(x,y) = g(x,y) = 0$, gives:

$$y = \frac{-ax - c}{b} = \frac{-Ax - C}{B} \tag{31}$$

which defines two linear functions. The intersection of these lines lies at the point:

$$x = \frac{Bc - bC}{Ab - aB}, \quad y = \frac{aC - Ac}{Ab - aB} \tag{32}$$

What remains is to evaluate the linear coefficients a, b, c, A, B, and C.

The implicit functions $f(x,y)$ and $g(x,y)$ can be thought of as three-dimensional surfaces above the x,y plane. For a small range of x and y, the surface is approximately planar. By finding three points on this plane, the plane is uniquely defined. So, two additional points near the original are introduced. This is done by calling the first point x_2, y_2. This point has corresponding values of f_2 and g_2. The second point is defined as $x_1 = x_2 + \Delta x$, $y_1 = y_2$, and has its corresponding values f_1 and g_1. The third point is $x_3 = x_2$, $y_3 = y_2 + \Delta y$, and has its values of f_3 and g_3. The value of Δx and Δy was set arbitrarily low at 0.00001. Then the actual values of the f and g functions are calculated for each of the three points using the numerical methods for Bessel/modified Bessel functions described in equations 20 - 23. From these, the linear coefficients can be calculated from equations 29 and 30, such that:

$$a = \frac{f_1 - f_2}{\Delta x} \tag{33}$$

$$b = \frac{f_3 - f_2}{\Delta y} \tag{34}$$

$$c = f_2 - ax_2 - by_2 \tag{35}$$

$$A = \frac{g_1 - g_2}{\Delta x} \tag{36}$$

$$B = \frac{g_3 - g_2}{\Delta y} \tag{37}$$

$$C = g_2 - Ax_2 - By_2 \tag{38}$$

These six coefficients are then used to generate a new point x_2, y_2 from equation 32, and the procedure re-iterates until the simultaneous equations are satisfied, i.e. $f(x_2,y_2) = g(x_2,y_2) = 0$. Once the solution has been reached and values for x ($x = k_{\rho 1}a$), y ($y = k_o a$), and z ($z = k_{\rho 2}a$) are known, the dielectric constant of the resonator, ε_{r6}, can be calculated from equation 9:

$$\varepsilon_{r6} = \frac{z^2 + x^2}{y^2} + 1 \tag{39}$$

Dielectric constants calculated in this way have a typical error of ~+15%.

An improvement to the model can be made by accounting for the fields in regions 3 and 5. So far the model assumes the fields to be zero here, creating a discontinuity in the field distribution on the boundaries of these regions. If one chooses instead appropriate electric fields for these regions, a continuous distribution can be restored; but the equations necessary do not satisfy the Helmholtz wave equation, making it impossible to construct a self-consistent magnetic field in these regions. In this case, the magnetic fields are simply left to be zero. To compensate for this case, artificial surface electric currents are introduced on the boundaries of regions 3 and 5, as shown in fig. 3. This technique is called a variational method, and the net effect is to lower the dielectric constant for a given resonant frequency.

The variational formula for computing the resonant frequency of the model in fig. 3 was derived by Harrington[8] using Rumsey's[9] reaction concept. The actual formula is a significantly non-trivial integral equation, but can be simplified into the form:

$$\omega_r^2 = \omega_0^2 \frac{D_1 + D_2 + N_3 + D_4 + N_5 + D_6 + N_h + N_v}{D_1 + D_2 + D_3 + D_4 + D_5 + D_6} \tag{40}$$

where ω_0 is the measured resonant frequency, ω_r is a reduced frequency, and the remaining terms are complicated functions of the fields and electrical parameters. The ErCalc[10] algorithm finishes by comparing this frequency to the one measured experimentally and adjusting ε_{r6} accordingly until $\omega_r = \omega_0$. The process can take several iterations to converge at a solution, which is why a computer is essential in the calculations! The results thus obtained show a typical error of ~-6%.

In code previously published[4] in the 1980s to calculate resonant frequency of dielectrics with known permittivities, none of the various Bessel functions were actually calculated; instead, approximate curve fits (equations 41 and 42) for the appropriate ratio of these functions were used, introducing a small but measurable error (figs. 8 - 9) in the region of interest.

Programming compilers capable of handling the very large and very small numbers involved in these calculations with sufficient accuracy did not exist then; however, such limitations no longer apply for arguments in the range of interest for most resonance tests.

$$\frac{J_0(z)}{J_1(z)} = \frac{0.0282z^5 - 0.4568z^4 + 3.0201z^3 - 10.5766z^2 + 20.5954z - 17.3551}{z - 3.8317} \tag{41}$$

$$\frac{K_0(z)}{K_1(z)} = \frac{z^5}{z^5 + 0.49907z^4 - 0.11226z^3 + 0.06539z^2 - 0.02679z + 0.00445} \tag{42}$$

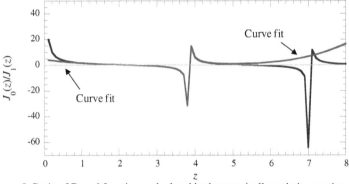

Figure 8. Ratio of Bessel functions calculated both numerically and via equation 41

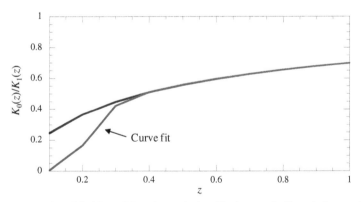

Figure 9. Ratio of modified Bessel functions calculated both numerically and via equation 42

The errors shown at the extremes in figs. 8 and 9 are large but not very relevant, as they correspond to geometries which are impossible to achieve in practice. The smaller error in the usable range of figs. 8 and 9 can be quantified for a specific case by making L_1 and L_2 very small (e.g., $L_1 = L_2 = 1 \times 10^{-7}$ mm) and comparing the results to that obtained in the more precise geometry of Hakki and Coleman,[2] where $L_1 = L_2 = 0$. In such a geometry, for example, a typical resonator with, say, $\varepsilon_r = 75.97701$, diameter 6 mm and thickness 3 mm has a resonant frequency of 8 GHz. The error in ε_r calculated by using ratios is +0.138%. The error when numerically calculating each Bessel function is only -0.025%. Obviously, these errors seem very small, but are necessarily scaled up as L_1 and L_2 increase. If $\varepsilon_r(var)'$ is the variationally improved dielectric constant calculated by ratios and $\varepsilon_r(var)''$ is that obtained by calculating each Bessel function, then typically $\varepsilon_r(var)' = \varepsilon_r(var)'' \pm 4$. Clearly, the improvement in accuracy is important; and with modern computational abilities, the increased time required for the full calculation is almost negligible. In the ErCalc algorithm, although the numerical calculation of $K_0(z)$ breaks down for $z > 10$ due to overflow errors, the curve fit (equation 42) is invoked in such cases in order to extend the usable range of the programme. The mean square error in equation 42 for $10 < z < 12$ is just 1.77×10^{-10}, so the resulting loss of accuracy is negligible.

CONCLUSION

Two programmes, HakCol and ErCalc, have been developed in order to solve for the relative permittivity of resonators in either the geometry of Hakki and Coleman[2] (i.e., Courtney[3]) or Itoh and Rudokas[7] as modified by Kajfez.[4] Both invoke numerical calculations of Bessel/modified Bessel functions to achieve highly accurate results.

REFERENCES
[1]O.V. Karpova, On an Absolute Method of Measurement of Dielectric Properties of a Solid Using a Π-Shaped Resonator, *Sov. Phys.*, **1**, 220-228 (1959).
[2]B.W. Hakki and P.D. Coleman, A Dieelctric Resonator Method of Measuring Inductive Capacities in the Millimeter Range, *IRE Trans. Microwave Theory Tech.*, **MTT-8**, 402-410 (1960).
[3]W.E. Courtney, Analysis and Evaluation of a Method of Measuring the Complex Permittivity and Permeability of Microwave Insulators, *IEEE Trans. Microwave Theory Tech.*, **MTT-18**, 476-485 (1970).
[4]D. Kajfez and P. Guillon, *Dielectric Resonators*, Artech House, Inc., Dedham, MA (1986).
[5]S.B. Cohn and K.C. Kelly, Microwave Measurement of High-Dielectric Constant Materials, *IEEE Trans. Microwave Theory Tech.*, **MTT-14**, 406-410 (1966).
[6]http://coen.boisestate.edu/rickubic/files/2012/05/HakCol.exe
[7]T. Itoh and R.S. Rudokas, New Method for Computing the Resonant Frequencies of Dielectric Resonators, *IEEE Trans Microwave Theory Tech.*, **MTT-25**, 52-54 (1977).
[8]R.F. Harrington, *Time-Harmonic Electromagnetic Fields*, McGraw-Hill, New York (1961).
[9]V.H. Rumsey, The Reaction Concept in Electromagnetic Theory, *Phys. Rev., Series 2*, **94**, 1483-91 (1954).
[10]http://coen.boisestate.edu/rickubic/files/2012/05/ErCalc.exe

EFFECTS OF MgO ADDITIVE ON STRUCTURAL, DIELECTRIC PROPERTIES AND BREAKDOWN STRENGTH OF Mg$_2$TiO$_4$ CERAMICS DOPED WITH ZnO-B$_2$O$_3$ GLASS

Xiaohong Wang[a,b], Mengjie Wang[a,b], Zhaoqiang Li[a,b], Wenzhong Lu[a,b]

[a] School of Optical and Electronic Information, Huazhong University of Science and Technology, Wuhan 430074, China
[b] Key Lab of Functional Materials for Electronic Information (B), MOE, Wuhan 430074, China

ABSTRACT

The effects of MgO on the microwave dielectric properties and the breakdown strength of Mg$_2$TiO$_4$ ceramics doped with ZnO-B$_2$O$_3$ glass sintered at temperature from 1250°C to 1390°C have been investigated. When 1wt% Zn-B glass was added to xMgO-TiO$_2$ (x=1.5, 2, 2.5, 3, 3.5) ceramics, the sintering temperature of xMgO-TiO$_2$ ceramics was reduced from 1450°C to 1300°C. The microwave dielectric properties and the breakdown strength of the ceramics were strongly related to the MgO content. The optimized microwave dielectric properties with ε_r =13.15, $Q \times f$=139522 GHz, τ_f=-64.80 ppm/°C were achieved for 3.5MgO-TiO$_2$-1wt%Zn-B sintered at 1340°C for 6h. The best breakdown strength value (E_b=35.65 kV/mm) was achieved for 3.0MgO-TiO$_2$-1wt%Zn-B ceramics.

1. INTRODUCTION

In the development of modern wireless communication, materials with a low dielectric loss (high quality factor Q=1/tanδ) in the microwave range are used. The developments of microwave dielectrics based on the MgO-TiO$_2$ system have brought much more attention for its high-Q and low cost. In this binary system, there are three crystal phases Mg$_2$TiO$_4$, MgTiO$_3$ and MgTi$_2$O$_5$. Among these compositions, Mg$_2$TiO$_4$ and MgTiO$_3$ are characterized by a negative coefficient of the resonant frequency (τ_f=-50ppm/°C), dielectric constant ε_r=14 and 17, and high value of quality factor $Q \cdot f$= 150000GHz and 160000GHz, respectively[1-4], which are good candidates for use in microwave dielectrics. Meanwhile, MgTi$_2$O$_5$ with high dielectric loss ($Q \cdot f$= 47000GHz) [5] is thermodynamically more stable than MgTiO$_3$ and usually formed as an intermediate phase in the MgTiO$_3$ ceramic synthesised by the conventional solid-state reaction method [6].

Up until now, the main focus of MgO-TiO$_2$ systems has been on the investigation of dielectric loss and the coefficient of resonate frequency for wireless communication applications, few works have been devoted to the study of bulk breakdown strength (E_b) of MgO-TiO$_2$ composites. However, the bulk breakdown strength is an important electrical parameter for insulating materials. For example, the Dielectric Wall Accelerator, which is an approach that permits a ten-fold improvement in the performance of particle accelerators, requires insulating materials with high breakdown strength between electrodes in configurations that permit the greatest possible electric field gradients.

In this present work, the roles of MgO in ZnO-B$_2$O$_3$ fruit doped MgO-TiO$_2$ system were developed based on the following considerations: First, Belous[4] reported the addition of the dopant ZnO-B$_2$O$_3$ in Co-doped Mg$_2$TiO$_4$ ceramic resulted in a significant (up to 150° -200°C) reduction of the sintering temperature and only a slight decrease in the $Q \cdot f$ value, which was due to the effect of MgO-rich inclusions in the matrix phase. K. Sreedhar[7] reported an excess MgO could facilitate the conversion of $MgTi_2O_5$ to $MgTiO_3$ phase. Secondly, MgO has a high E_b (~100kV/mm)[8] and a low dielectric loss, the combined Mg_2TiO_4/MgO or $MgTiO_3$/MgO dielectric composites should have the ability to bear a much higher E_b. Finally, MgO could act as a grain growth inhibitor and make the microstructure more uniform [9]. In this paper, xMgO-TiO_2 system (x=1.5, 2, 2.5, 3, 3.5) doped with ZnO-B_2O_3 glass were prepared by a conventional mixed-oxide route. The phase composite, microstructure, microwave dielectric properties and breakdown strength of MgO-TiO_2 ceramic system were investigated accordingly.

2. EXPERIMENTAL PROCEDURE

The MgO-TiO_2 ceramics with MgO and TiO_2 in a different molar ratio of x (x=1.5, 2, 2.5, 3, 3.5) were prepared by the conventional solid state reaction method from high-purity $MgCO_3$ and TiO_2. Stoichiometric quantities of starting materials according to the compositions were mixed in a ball mill in ethanol for 3 h. After drying and sieving, the mixtures were heated at 1190 °C in air for 3 h. For the glass preparation, the reagent grade ZnO and H_3BO_3 precursors were appropriately weighted in a molar ratio of 1:2 and mixed. The mixture was then melted at 1100 °C for 2 h in an alumina crucible and quenched at room temperature in deionized water. This glass was roughly crushed in an alumina mortar and then grinded in deionized water in a planetary for 3 h to obtain fine powder. After drying and sieving, the glass frit powder (ZB) was prepared. The MgO-TiO_2 powders were ball-milled together with the 1.0 wt.% ZB glass frits in alcohol for 3 h. After drying at 70° -90°C, the powders were granulated with 5wt% polyvinyl alcohol (PVA) and then pressed to form green pellets with a diameter of 15mm, a height of 7-9 mm and 1-2mm separately, which were then sintered at 1200° -1390°C for 6 h. Some sintered samples were polished to 1mm in thickness and electrode with silver paint for the measurement of E_b.

The phase compositions of MgO-TiO_2 system were identified by the X-ray diffraction (XRD: Shimadzu XRD-7000, Japan). The microstructures and the phase distributions of the samples were examined by means of field-emission scanning electron microscope (FE-SEM: Quanta 200, Holland) using energy dispersive X-ray spectroscopy (EDX). Density measurement was performed using the Archimedes method. The dielectric constants (ε_r) and quality factors (Q·f) at frequencies around 7.5GHz were measured by the Hakki-Coleman method [10] using a vector network analyzer (ADVENT R3767C, Japan). The resonant frequency was measured in the temperature range of 25° -75°C and τ_f was calculated from the following equation:

$$\tau_f = \frac{(f_{75} - f_{25}) \times 10^6}{f_{25} \times 50}$$

(1)

where f_{75} and f_{25} are the resonant frequencies at 75°C and 25°C respectively. The DC breakdown measurement was performed using a breakdown voltage tester (HJC-60KV, China) at room temperature. All the samples were immersed in transformer oil to prevent surface flashover.

3. RESULTS AND DISCUSSION

Fig. 1 shows the XRD patterns of x MgO-TiO$_2$ ceramics doped with 1 wt% ZnO-B$_2$O$_3$ frit sintered at 1340°C. Mg$_2$TiO$_4$ phase (ICDD-PDF#00-025-1157) is detected in all samples. The MgTiO$_3$ phase (ICDD-PDF#00-006-0494) is found to coexist along with the major phase of Mg$_2$TiO$_4$ for the sintered 1.5MgO-TiO$_2$ mixture. And MgTiO$_3$ phase is not detected as MgO content increased to $x \geq 2$ in xMgO-TiO$_2$ ceramics. Petrova et al [11] have reported that the thermal decomposition of Mg$_2$TiO$_4$, which accompanied by the formation of MgTiO$_3$ in the following way: Mg$_2$TiO$_4 \rightarrow$ Mg$_2$TiO$_4$+ MgTiO$_3$, becomes negligible when the temperature exceeds 1400°C. However, the XRD analysis indicates that single-phase Mg$_2$TiO$_4$ is formed at 1340°C in 1wt%ZB glass doped 2MgO-TiO$_2$ ceramic. This fact has also been confirmed by a detailed microstructural analysis (Fig. 2(a)). It implies that the thermal decomposition is prevented, which might be due to the addition of ZB frit as sintering aids. Although the strongest diffraction peaks of MgO (ICDD-PDF#01-075-0447) is overlapped by peaks of Mg$_2$TiO$_4$, the intensity of the peak around 42.8° enhances gradually with the increasing of x, it implies that the excess MgO presents in the samples at a higher MgO concentration ($2.5 \leq x \leq 3.5$).

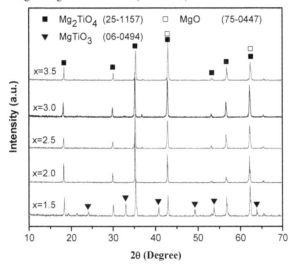

Fig.1 XRD patterns of xMgO-TiO$_2$ ceramics with 1 wt% ZnO-B$_2$O$_3$ frit

Backscattered SEM micrographs of sintered ceramic surfaces are shown in Fig.2. Energy dispersive X-ray spectroscopy (EDX) was taken for the composition analysis. The EDX spectra and composition of the spot A-F in Fig.2 are shown in Fig.3. All SEM images show dense microstructures with a low porosity. The combined analysis of SEM and EDX indicates that MgTiO$_3$ is presented in 1.5MgO-TiO$_2$ ceramic; meanwhile the dark grains in Fig.2 (c)-(e) are identified as MgO. In the Fig.2 (b), the compositions of the large and small grains are identified as Mg$_2$TiO$_4$. With the increasing of MgO content, the Mg$_2$TiO$_4$ grain size distribution becomes more homogeneous.

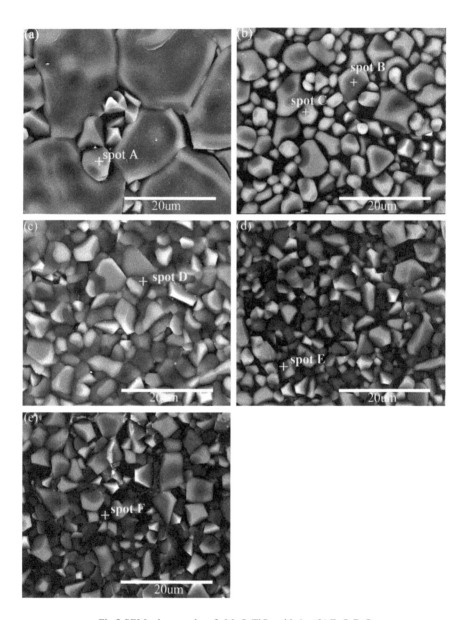

Fig.2 SEM micrographs of xMgO-TiO$_2$ with 1 wt% ZnO-B$_2$O$_3$
(a) x=1.5; (b) x=2.0; (c) x=2.5; (d) x=3.0; (e) x=3.5

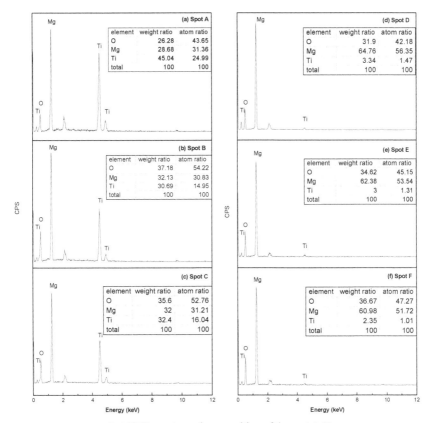

Fig.3 EDX spectra and composition of the spot A-F

Variation of density with sintering temperature for xMgO-TiO$_2$ with 1 wt% ZnO-B$_2$O$_3$ ceramics is shown in Fig.4. When 1wt% ZnO-B$_2$O$_3$ frit was added to the system xMgO-TiO$_2$, the sintering temperature decreased by 150°C: in this case, well sintered ceramics were formed at 1300 °C. Addition of low melting glasses enhances the sinterability of the ceramic powders due to the liquid phase sintering. In addition, the density reached the maximum at x=1.5. It was due to the appearance of MgTiO$_3$ phase with theoretical density of 3.89g/cm^3, which is higher than the theoretical density of MgO (3.59g/cm^3) and Mg$_2$TiO$_4$ (3.58g/cm^3). As the amount of MgO exceeded 2 wt%, the bulk density increases with x.

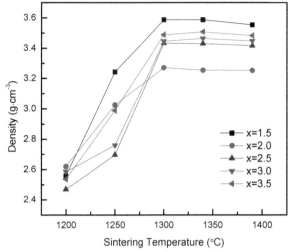

Fig.4 Variation of density with sintering temperature of xMgO-TiO$_2$ with 1 wt% ZnO-B$_2$O$_3$ frit

Fig.5 shows the microwave dielectric properties of 1 wt% ZnO-B$_2$O$_3$ frit doped xMgO-TiO$_2$ ceramics as a function of the MgO content and sintering temperature. The correlations between ε_r value and sintering temperatures nearly reveal the same trend as those between densities and sintering temperatures, as observed in Fig.4 and Fig.5(a). The maximum ε_r value can be achieved at 1300°C for x=1.5. Furthermore, under the same sintering condition, with increasing MgO content, ε_r values decrease substantially and then increase a bit at $2\leq x\geq 2.5$, finally decrease when $x\geq 2.5$, the tendencies of which are partially in accordance with the variation of densities. It is suggested that the MgO content is also an important factor to control dielectric constants which is consistent with the logarithmic mixing rule [12]. As the dielectric changes of MgTiO$_3$ (ε_r=17) is higher than those of Mg$_2$TiO$_4$ (ε_r=14) and MgO (ε_r~9.8), the formation of MgTiO$_3$ phase in 1.5MgO-TiO$_2$ ceramics improves ε_r value, meanwhile the dielectric constant decreases with appearance of MgO phase. The low ε_r value is achieved at x=2, which is mainly attributed to the high porosity.

Fig.5 (b) illustrates the $Q{\cdot}f$ dependence of sintering temperature for 1 wt% ZB frit doped xMgO-TiO$_2$ ceramics. For a given x value, the $Q{\cdot}f$ value initially increases and then decreases after reaching the maximum at 1300°C for $2\leq x\leq 3$ and 1340 °C for x=1.5, 3.5. It is believed that the densities play an important role in controlling dielectric loss, as has been often found in other microwave dielectric materials [13, 14]. The $Q{\cdot}f$ value is generally affected not only by the lattice vibrational mode, but also the pores, the second phase, the impurities, the lattice defect, crystallizability and inner stress. The increase in sintering temperature is beneficial to the densification and crystallizability until the $Q{\cdot}f$ value reach the maximum. The further increase in sintering temperature will result in the appearance of abnormal grains and pores and consequently lead to the reduction of the $Q{\cdot}f$ value. Furthermore, the maximums of the $Q{\cdot}f$ value are promoted with MgO content when $x\geq 2.0$ due to homogeneous grains distribution caused by MgO addition and the increases in the second phase (MgO) with high $Q{\cdot}f$ value (751500 GHz for

0.96MgO-0.04LiF) [15]. A maximum $Q \cdot f$ value of 139522 GHz is obtained for the 1wt% ZB frit doped 3.5MgO-TiO$_2$ ceramics sintered at 1340 °C.

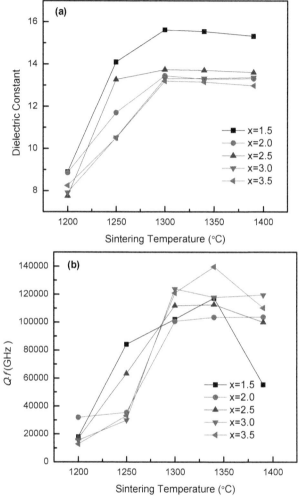

Fig. 5 Microwave dielectric properties of x MgO-TiO$_2$ ceramics with 1wt% ZnO-B$_2$O$_3$ frit

Fig.6 illustrates the temperature coefficients of resonant frequency (τ_f) of 1wt% ZB frit doped xMgO-TiO$_2$ ceramics sintered at 1340 °C. The temperature coefficient of resonant frequency is well known to be governed by the composition, the additives and the second phase of the materials. Significant variation in the τ_f value is not observed for specimens with different content of MgO due to the τ_f value of MgO, MgTiO$_3$ and Mg$_2$TiO$_4$ is similar.

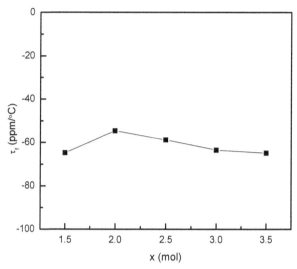

Fig.6 τ_f of xMgO-TiO$_2$ ceramics with1wt% ZnO-B$_2$O$_3$ frit sintered at 1340 °C

The average breakdown strength (E_b) of 1 wt% ZnO-B$_2$O$_3$ frit doped xMgO-TiO$_2$ ceramics is shown in Fig.7. The breakdown strength of the ceramics depends not only on the porosity, uniformity and grain size of the materials, but also on the testing condition, such as sample thickness, electrode geometry and area[16-20]. In this study, all of the samples for the breakdown strength testing were polished to 1±0.02mm in diameter and the electrode was 11 mm in diameter. The correlations between E_b and sintering temperatures reveal the same trend as those between densities and sintering temperatures, as observed in Fig.4 and Fig. 7. Pure Mg$_2$TiO$_4$ ceramics shows a maximum E_b of 24.82kV/mm at 1300 °C, while the maximum E_b can be achieved at 1340 °C for the others. As it can be seen from the Fig.7, with the increase of x, the average E_b decreases to a minimum value of 19.26kV/mm at x=2 and then increases substantially to a maximum value at x=3, finally decreases of xMgO-TiO$_2$ ceramics sintered at 1340 °C. The highest E_b of 35.65kV/mm is obtained for composition of 3MgO-TiO$_2$, which is about 1.43 times that of the pure Mg$_2$TiO$_4$ ceramics. Gerson [16] discussed the effect of porosity in ceramic materials on their dielectric breakdown strength by a statistical approach. The theory shows that the E_b decreases with the increase of porosity and pore size. Beauchamp [17] and Tunkasiri [18] reported the reduction of the grain size and a more homogeneous microstructure is beneficial to the improvement of E_b. In this present work, the porosity and grain size of the samples could be reduced with the addition of MgO, which is believed to be the main reason for the improvement of E_b.

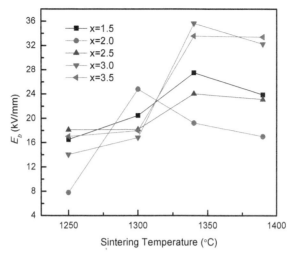

Fig. 7 Breakdown strength of xMgO-TiO$_2$ ceramics with 1wt% ZnO-B$_2$O$_3$ sintered at 1340 °C

4. CONCLUSIONS

(1) In the system xMgO-TiO$_2$(1.5≤x≤3.5) containing the Zn-B frit additive, the Mg$_2$TiO$_4$ phase is formed in all of the composites, while MgTiO$_3$ and MgO appears in x=1.5 and 2<x≤3.5 respectively.

(2) The introduction of 1 wt% of the dopant ZnO-B$_2$O$_3$ frit into the xMgO-TiO$_2$ ceramics results in a 150 °C reduction in the sintering temperature and prevents the thermal decomposition of Mg$_2$TiO$_4$ at 1340 °C.

(3) The appearance of MgO phase in xMgO-TiO$_2$ restrains the grain growth and makes the microstructure more uniform.

(4) The microwave dielectric properties and the breakdown strength of the ceramics were strongly related to the MgO content in the system xMgO-TiO$_2$(1.5≤x≤3.5) containing 1wt% ZnO-B$_2$O$_3$ frit additive. The optimized microwave dielectric properties with ε_r =13.15, $Q \times f$=139522 GHz, τ_f=-64.80 ppm/°C were achieved for 3.5MgO-TiO$_2$ sintered at 1340°C, while the highest E_b of 35.65 kV/mm is obtained in the 3.0MgO-TiO$_2$ composites.

ACKNOWLEDGMENTS

This work was supported by the Fundamental Research Funds for the Central Universities of China (2012QN153). The authors are grateful to Analytical and Testing Center, Huazhong University of Science and Technology for XRD and SEM analysis.

REFERENCES

[1] J. H. Sohn, Y. Inaguma, S. O. Yoon, M. Itoh, T. Nakamura, S. J. Yoon, and H. J. Kim, Microwave Dielectric Characteristic of Ilmenite-type Titanates with High-Q Values, *Jpn. J. Appl. Phys.*, **33**, 5466-70, (1994)

[2] C. L. Huang, C. L. Pan, Low-temperature Sintering and Microwave Dielectric Properties of (1-x)MgTiO$_3$-xCaTiO$_3$ Ceramics Using Bismuth Addition, *Jpn. J. Appl. Phys.*, **41**, 707-11, (2002)

[3] K. Wakino, Recent Developments of Dielectric Resonator Materials and Filters, *Ferroelectrics*, **91**, 69-86, (1989)

[4] A. Belous, O. Ovchar, D. Durilin, M. M. Krzmanc, M. Valant, D. Suvorov, High-Q Microwave Dielectric Materials Based on the Spinel Mg$_2$TiO$_4$, *J. Am. Ceram. Soc.*, **89**, 3441-45, (2006)

[5] Hyunho Shin, Hee-Kyun Shin, Hyun Suk Jung, Seo-Yong Cho, Kug Sun Hong, Phase Evolution and Dielectric Properties of MgTi$_2$O$_5$ Ceramic Sintered With Lithium Borosilicate Glass, *Mater. Res. Bull.*, **40**, 2021-28, (2005)

[6] X. H. Zhou, Y. Yuan, L. C. Xiang, Y. Huang, Synthesis of MgTiO$_3$ By Solid State Reaction and Characteristics With Addition, *J. Mater. Sci.*, **42**, 6628-32, (2007)

[7] K. Sreedhar, N. R. Pavaskar, Synthesis of MgTiO$_3$ and Mg$_4$Nb$_2$O$_9$ Using Stoichiometrically Excess MgO, *Mater. Lett.*, **53**, 452-5, (2002)

[8] E. K. Beauchamp, Effect of Microstructure on Pulse Electrical Strength of MgO, *J. Am. Ceram. Soc.*, **54**, 484-7, (1971)

[9] J. S. Park and Y. H. Han, Effects of MgO Coating on Microstructure and Dielectric Properties of BaTiO$_3$, *J. Euro. Ceram. Soc.*, **27**, 1077-82, (2007)

[10] B. W. Hakki, P. D. Coleman, A Dielectric Resonator Method of Measuring Inductive Capacities in the Millmeter Range, *IRE Trans MTT*, **8**, 402-10, (1960)

[11] M. A. Petrova, G. A. Mikirticheva, A. S. Novikova, and V. F. Popova, Spinel Solid Solutions in the Systems MgAl$_2$O$_4$-ZnAl$_2$O$_4$ and MgAl$_2$O$_4$-Mg$_2$TiO$_4$, *J. Mater. Res.*, **12**, 2584-88, (1997)

[12] K.P. Surendran, N. Santha, P. Mohanan, M.T. Sebastian, Temperature Stable Low Loss Ceramic Dielectric in (1-x)ZnAl$_2$O$_4$-xTiO$_2$ System for Microwave Substrate Applications, *Eur. Phys. J. B*, **41**, 301-6, (2004)

[13] Y. Zheng, X. Zhao, W. Lei, S. X. Wang, Effect of Bi$_2$O$_3$ Addition on the Microstructure and Microwave Dielectric Characteristics of Ba$_{6-3x}$(Sm$_{0.2}$Nd$_{0.8}$)$_{8+2x}$Ti$_{18}$O$_{54}$ (x=2/3) Ceramics, *Mater. Lett.*, **60**, 459-63, (2006)

[14] C. L. Huang, C. S. Hsu, R. J. Lin, Improved High-Q Microwave Dielectric Resonator Using ZnO and WO$_3$-doped Zr$_{0.8}$Sn$_{0.2}$TiO$_4$ Ceramics, *Mater. Res. Bull.*, 2001, **36**, 1985-93, (2001)

[15] A. Kan, T. Moriyama, S. Takahashi, and H. Ogawa, Low-Temperature Sintering and Microwave Dielectric Properties of MgO Ceramic with LiF Addition. *Jpn. J. Appl. Phys.*, **50**, 09NF02-1-5, (2011)

[16] R. Gerson and T. Marshall, Dielectric Breakdown of Porous Ceramics, *J. Appl. Phys.*, **30**, 1650-53, (1959)

[17] E. K. Beauchamp, Effect of Microstructure on Pulse Strength of MgO, *J. Am. Ceram. Soc.*, **54**, 484-87, (1971)

[18] T. Tunkasiri and G. Rujijanagul, Dielectric Strength of Fine Grained Barium Titanate Ceramics, *J. Mater. Sci. Lett.*, **15**, 1767-69, (1996)

[19] B. Gilmore, Development of High Energy Density Dielectrics for Pulsed Power Application, Ph.D. Dissertation, University of Missouri-Rolla, 2001

[20] I. O. Owate, R. Freer, Dielectric Breakdown of Ceramics and Glass Ceramics, *IEEE, in Proc. 6th Intl Conf. on Dielectric Materials, Measurements and Applications,* 443-46, (1992)

DESIGN OF MICROWAVE DIELECTRICS BASED ON CRYSTALLOGRAPHY

Hitoshi OHSATO

Nagoya Industrial Science Research Institute, Nagoya 464-0819, Japan
Nagoya Institute of Technology, Nagoya 466-8555, Japan

ABSTRACT

The authors have been studying correlation between crystal structure and microwave dielectric properties based on crystallography. New dielectric materials were designed based on the origins of the properties clarified as presented in following three categories. (I) Low loss microwave dielectrics designed by low internal strain due to compositional ordering, perfect crystallinity without defects and impurities, and high symmetry and high crystallographic densities. (II) Dielectric constants due to large unit cell with inversion symmetry i, large rattling factor accompanying expanding polyhedron. On the other hand, low dielectric constant due to tight polyhedron due to covalence such as silicates. (III) Low temperature coefficients of resonant frequencies (TCf) are affected by the tilted octahedra depending on the crystal transitions. The TCf is designed generally by combination positive and negative TCf, which might be clarified the origin of TCf. In this paper, some examples with relationship between crystal structure and microwave dielectric properties are presented, and origin of the properties are clarified.

INTRODUCTION

The authors have been studying correlation between crystal structure and microwave dielectric properties based on crystallography[1-3]. Microwave dielectrics are expected following properties: high quality factor Q for resonate with microwave, dielectric constant ε_r of high for shortage wavelength, and low for millimeter wave, and near zero temperature coefficient of resonate frequency TCf for stability usage on wide temperature range. These properties are depending on the crystal structure. So, after relationship between crystal structure and properties has been studied and the origin of the properties is clarified, new dielectrics with high properties have been designed.

New dielectric materials improved the properties were designed based on the origins of the properties clarified as presented in following three categories. (I) Low loss microwave dielectrics designed by low internal strain due to compositional ordering, perfect crystallinity without defects and impurities, and high symmetry and high crystallographic densities. (II) High dielectric constants due to large unit cell with inversion symmetry i, accompanying large rattling factor on the expanding polyhedron. On the other hand, low dielectric constant due to tight polyhedron due to covalence such as silicates. (III) Near zero temperature coefficient of resonant frequency TCf designed generally by combination positive and negative TCf which has clarified the origin of *TCf*.

In this paper, examples with relationship between crystal structure and microwave dielectric properties of the microwave dielectrics are presented, and origins of the properties are clarified.

EXPERIMENTAL

These compounds are almost fabricated by solid state reactions, identified and obtained the lattice parameters by X-ray powder diffraction (XRPD)[4]. The crystal structures were analyzed by RADY program[5] for single crystal, or by Rietveld method[6] for powder patterns. And the microwave dielectric properties were evaluated by Hakki and Colleman's method[7-8], as presented previous papers[1-2].

RESULTS AND DISCUSSIONS

Three microwave dielectric properties are presented in the order of High Q, ε_r and *TCf*.

I) High Q

I-1) (a) High Q by compositional ordering

The pseudo-tungstenbronze $Ba_{6-3x}R_{8+2x}Ti_{18}O_{54}$ (R = rare earth) solid solutions shows the highest Qf value at $x = 2/3$ as shown in Fig. 1(a)[1,9], which internal strain is the smallest at the composition as shown in Fig. 1(b).

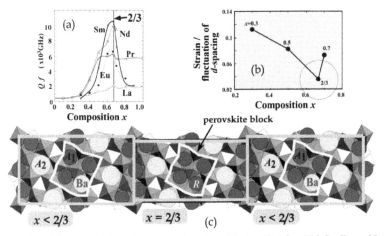

Fig. 1. (a) The Qf of pseudo-tungstenbronze $Ba_{6-3x}R_{8+2x}Ti_{18}O_{54}$ (R = Sm, Nd, Pr, Eu and La) as a function of composition x. (b) Strain/fluctuation of d-spacing of the Sm-analogous as a function of x. (c) Crystal structure with different occupation in the perovskite blocks in $x < 2/3$ case.

At the $x = 2/3$ composition of the solid solutions, the crystal structure show compositional ordering for R ions occupying A_1 site in perovskite block as shown in a part of $x = 2/3$ on Fig. 1(c). At the $x < 2/3$, Ba ions occupy statistically at A_1 sites in the perovskite block such as $[R_{8+2x}Ba_{2-3x}V_x]_{A1}$ as shown in a part of $x < 2/3$ on Fig. 1(c). The internal strains as shown in Fig. 1(b) are explained as the fluctuation of d-spacing of lattice constants. At $x < 2/3$, the fluctuation of d-spacing becomes large depending on the statistical occupation of Ba ions in perovskite blocks. On the other hand, at $x = 2/3$, the fluctuation of d-spacing becomes to be reduced depending on the all unit cells with the same size.

(b) Design more compositional ordering on the pseudo-tungstenbronze compounds

Low Qf value of ca. 200 GHz on the $Ba_6R_8Ti_{18}O_{54}$ (R = Nd) composition in the vicinity of $x = 0$ for $Ba_{6-3x}R_{8+2x}Ti_{18}O_{54}$ as shown in Fig. 1(a) was improved to 6,000 GHz by substitution of Sr for Ba in the perovskite blocks of $[Nd_8Ba_{2-\alpha}Sr_\alpha]_{A1}[Ba_4]_{A2}Ti_{18}O_{54}$ as shown in Fig. 2[10]. The substitution of Sr with smaller ionic radius than Ba ion must reduce the internal strain decreasing the fluctuation of d-spacing. This substitution of Sr for Ba also introduces a kind of the compositional ordering in A_1 and A_2 sites.

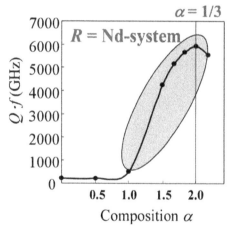

Fig. 2. Qf values of the Nd-analogy with $x = 0$ were improved from 200 to 6,000 GHz by substitution of Sr for Ba[10].

I-2) High Q by high symmetry

As presented at previous section, compositional ordering on the pseudo- tungsten bronze solid solutions brings high Q. And many researchers presented ordering of complex perovskite such as $Ba(Mg_{1/3}Ta_{2/3})O_3$ (BMT), $Ba(Zn_{1/3}Ta_{2/3})O_3$ (BZT) and $Ba(Zn_{1/3}Nb_{2/3})O_3$

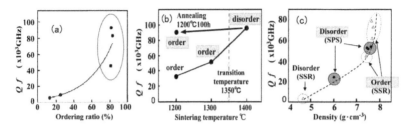

Fig. 3. (a) Qf of BZT ceramics as a function of ordering ratio, (b) Qf of BZN with order-disorder phase transition at 1350 °C as a function of sintering temperature, (c) Qf of BZT as a function of density by solid state reaction (SSR) and spark plasma sintering (SPS). Order: ordered perovskite, Disorder: disordered perovskite.

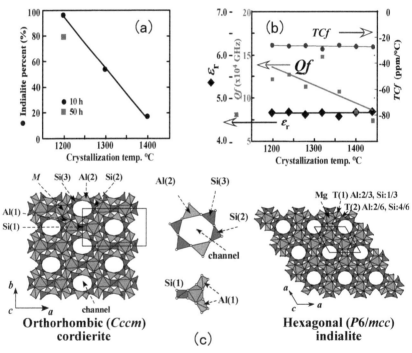

Fig. 4. (a) Indialite percentage sintered for 10 h as a function of temperature, (b) Qf, ε_r and TCf of indialite/cordierite glass ceramics sintered at 1200 °C for 10 h as a function of crystallized temperature, (c) polymorph of indialite/cordierite: cordierite is ordered form with orthorhombic, and indialite is disorder form with hexagonal.

(BZN) brings high $Q^{11-12)}$. On the other hand, Koga et al.[13-17] found another factor instead of ordering, and Ohsato et al. presented high symmetry brings high Q instead of ordering on the compounds with order-disorder transition[18-20]. This conclusion has been derived from following five examples: (1) in the case of BZT, the Q values are not depended on the ordering ratio[13-15] as shown in Fig. 3(a). (2) In the case of BZN[17] with an clear order-disorder transition at 1350 °C, the low temperature form with high ordering ratio did not show higher Q than the high temperature form with high symmetry as shown in Fig. 3(b). (3) The disordered BZT samples synthesized by spark plasma sintering (SPS) showed the same high Q as ordered ones synthesized by solid state reaction (SSR)[16] as shown in Fig. 3(c). (4) Ni-doped cordierite[21-23] changing to disordered high temperature form was improved in the Q value. (5) In the case of indialite/cordierite glass ceramics, indialite with high symmetry ($P6/mcc$) shown higher Q than cordierite (Ccm) as shown in Fig. 4. Si/AlO$_4$ tetrahedra of indialite and cordierite are disordered and ordered, respectively[24].

The authors resume about the order of crystal structure and microwave dielectric properties. There are two categories of order: one is compositional ordering without order-disorder transition, another is ordering with order-disorder transition. The former case is pseudo-tungstenbronze solid solutions, the latter case is complex perovskite such as BMT, BZT and BZN, as presented above. The ordering of former yield on the same crystal symmetry, but that of the latter on the different symmetry, that is, ordered complex perovskite on the hexagonal and disordered one on the cubic. In the case of complex perovskite with A-site ordering, effect of symmetry might be dominant more than that of ordering.

I-3) High Q by perfect crystal structure

As an example of microwave dielectrics with defect, complex perovskite BZT on the BaO side in the vicinity of BZT as shown in Fig. 5(a) is presented here[15-16]. The phase is a single solid solution with disordered structure and has defects in B- and O-sites, as presented by Koga et al. The Qf values of the side become low in order of A, Q, R and S on the ⓢ line as shown in Fig. 5(b). Kugimiya[25] also presented a defect phase in the region $\alpha > 5g/4$ in Ba$_\alpha$Ta$_g$O$_{\alpha+5g/2}$ as shown in Fig. 5(c), the composition denoted by Ba$_{1+\alpha}$(Mg$_{1/3}$Ta$_{2/3+\gamma}$V$_{\alpha-\gamma}$)O$_{3+\alpha+5\gamma/2}$V$_{2\alpha-5\gamma/2}$ has B- and O-site vacancies with holes and electrons.

Though the crystal structural origins of Q factors stated above are intrinsic, extrinsic origin such as impurities, grain growth also degregate the Q values. The forsterite ceramics are improved from 10,000 to 240,000 GHz by means of using high purity raw materials as shown in Fig.6(a). The grains of forsterite are very clear and there is no glassy phase among the grains as shown in Fig. 7(c) [26-29].

Grain growth without rough microstructure sometimes improves Q values as shown in Fig. 7, which shows Qf of Al$_2$O$_3$ as a function of grain size. Q of Al$_2$O$_3$ was improved from

335,000 to 680,000 GHz by grain growth[30]. The Qf value of single crystal of Al_2O_3 shows 1,890,000 GHz on $// c$-axis[31]. So, grain growth might be in the process to single crystal with superior Qf.

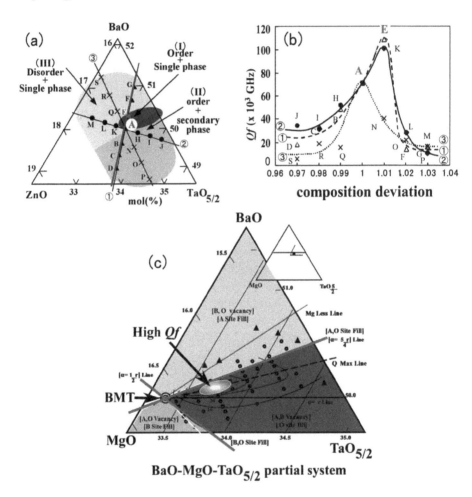

Fig. 5. (a) Partial ternary phase diagram around BZT. Three area are shown as (I) for order/single phase, (II) for order/secondary phase, (III) for disorder/single phase. (b) Qf as a function of composition deviation. On the line ③ , disordered single phase region (III) shows low Qf. (c) On the partial ternary system in the vicinity of BMT presented by Kugimiya. Highest Qf value is located near the line of BMT-BaTa$_{4/5}$O$_3$.

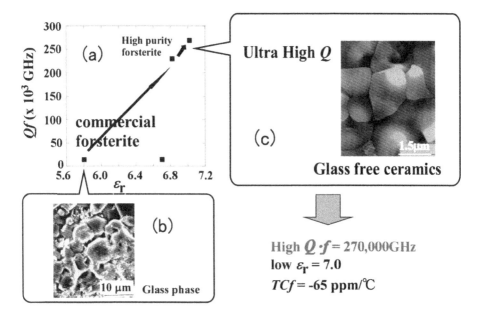

Fig. 6. (a) Qf improved by means of using high purity raw materials. (b) commercial forsterite with glass phase. (c) improved glass free forsterite with high Q.

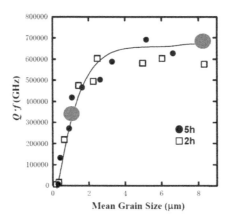

Fig. 7. Qf of Al_2O_3 as a function of mean grain size. Miyauchi improved the Qf of Al_2O_3 from 335,000 to 680,000 GHz.

II) Dielectric constant ε_r

Usually, high dielectric constant induces high dielectric loss, because the both properties are proportional to fluctuation of ions. So, Qf values decrease as a function of dielectric constants as shown in Fig. 8(a)[31-32]. Pseudo-tungstenbronze solid solutions with high Qf of ca. 10,000 GHz are examples for high dielectric constants of 80-90[34]. The dielectric constants are almost proportional to the unit cell volumes as shown in Fig. 8(b)[33]. This compound has large unit cell volume of over 2,000 Å^3 and 1/8 of the unit cell is an asymmetric unit, which space group is *Pnma* (No.62), and multiplicity/Wyckoff letter is 8d[1]. The large dielectric constants are produced from the 4.5 TiO$_6$ octahedra in an asymmetric unit as shown in Fig. 8(c).

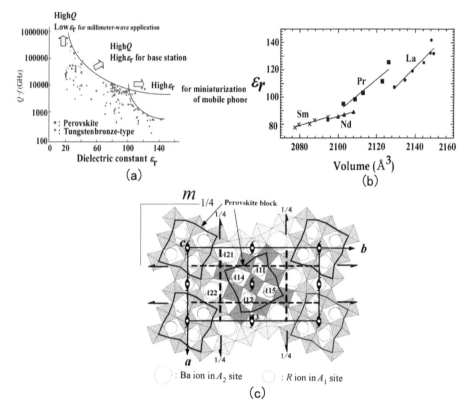

Fig. 8. (a) Qf as a function of dielectric constants, (b) ε_r as a function of unit cell volume of the pseudo-tungstenbronze. (c) Large asymmetric unit ($a/2 * b/2 * c/2$) of the pseudo-tungstenbronze.

The TiO_6 octahedra of the pseudo-tungstenbronze structure are tilting forming super structure of two times of c-axis[1]. The tilting affects to dielectric constant. Fig. 9 and Table 1 show schematic figures and angles of tilting from c-axis, respectively. The tilting angles are depending on the composition and R cations. As the composition x of 0.5 has much more cations in A_2 sites than $x = 0.7$, the tilting angle decreases and dielectric constant increases. TiO_6 octahedron reached the straight to the c-axes that is the tilting angle of $0°$ yields large dielectric constants.

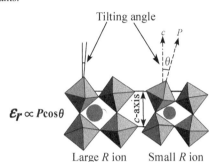

Fig. 9. Corelationship between dielectric constant and tilting angle.

Table 1. Tilting angles θ from c-axis on $x = 0.5$ and 0.7 for $Ba_{6-3x}R_{8+2x}Ti_{18}O_{54}$.

X	0.5	0.7
Sm	9.990°	10.63°
Nd	9.687°	8.961°

(a)

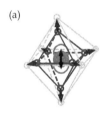

· TiO_6 octahedron is formed almost by ionic bond.

(b)

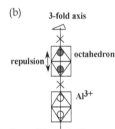

· Al ion located in octahedron, just on 3-fold axis and was fixed by the repulsion of each Al ion.

(c)

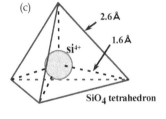

· SiO_4 tetrahdron is formed by ionic bond of 45% and covalent bond of 55%.

$$\varepsilon r_{titanate} > \varepsilon r_{aluminate} > \varepsilon r_{silicate}$$

Fig. 10. Dielectric constant due to crystal structure: TiO_6 octahedron (a), octahedra of Al_2O_3 (b), and SiO_4 tetrahedron (c).

Rattling of the cations also affects the dielectric constant. Fig. 10 compared with the effects of rattling on the titanate with TiO_6 octahedron, Al_2O_3 and silicates[35]. As the titanates have large rattling effect because of large space around Ti cation as shown in Fig. 10(a). As silicates composed by SiO_4 tetrahedron which combined a half by covalency, the cation of Si was bonded hard as shown in Fig. 10(c). So, as the rattling effects are reduced, the dielectric constant is small, which is suitable for millimeter wave dielectrics. The dielectric constant of rutile TiO_2 is about 10, although the structure composed by TiO_6 octahedra. As the octahedra of Al_2O_3 occupied by 2/3 by Ti atoms as shown in Fig. 10(b), Ti ions occupied two adjacent octahedra repulse to the corner of octahedron each other. So, the rutile shows medium dielectric constants.

III) Temperature coefficient of resonant frequency (*TCf*)

The *TCf* is defined as following equation:

$$TCf = (f_T\text{-}f_{ref})/f_{ref}(T\text{-}T_{ref}) \text{ ppm/}^{\circ}C \qquad (1)$$

Here, f_T and f_{ref} (GHz) are resonant frequencies on the temperature T, and the reference temperature T_{ref}, respectively. $T_{ref} = -40\ ^{\circ}C$ and $T = 85\ ^{\circ}C$ on JIS R 1627-1996[36]. The *TCf* is very difficult to estimate the value, but it is depending on the crystal structure. Reaney et al.[37] presented the relationship between temperature coefficient of dielectric constant (*TCε*) and tolerance factor t, given by eq. (2) on the complex perovskite as shown in Fig. 11[38].

$$t = (R_A + R_O)/\sqrt{2}\ (R_B + R_O) \qquad (2)$$

Here, R_A, R_B and R_O are ionic radii of A, B and O ions on ABO_3 perovskite, respectively. There is a relationship between *TCf* and *TCε* given by eq. (3).

$$TCf = -[\alpha + TC\varepsilon/2] \qquad (3)$$

Here, α is a coefficient of thermal expansion. The t is affected by tilting of octahedron: the octahedron is untilted, in the range of t between 1.055 and 0.985, antiphase tilted in 0.964 and 0.985, and in phase and antiphase tilted in 0.964 and 0.92 as shown in Fig. 11. BMT and $Sr(Mg_{1/3}Nb_{2/3})O_3$ (SMN) located near zero *TCε* at $t = 1.033$ and 0.964, respectively. Solid solutions such as $Ba_xSr_{1-x}(Mg_{1/3}Ta_{2/3})O_3$ (BSMT) also show the same trend. In the case of pseudo-tungstenbronze with tilted octahedra, *TCf* of Sm-compound is minus as opposite to Nd-, Pr-, and La-compounds with plus *TCf*. Sm-Nd- and Sm-La-pseudo-tungstenbronze solid solutions yield near zero *TCf*[39]. As seen in these examples, the resonant frequencies are affected by tilting of octahedra. On the other hand, usually, near zero *TCf* achieved adding different compound with opposite sign of *TCf*, and the adding ratio is in inverse proportion to the *TCf*. The *TCf* should be considered on the resonation which is affected by tilting of octahedron, and volume of additional compound for reducing of *TCf*.

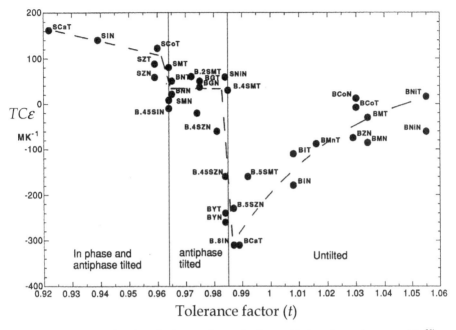

Fig. 11. The $TC\varepsilon$ as a function of tolerance factor t for Ba- and Sr-based complex perovskite[39].

SUMMARY

1)   Compositional ordering and high symmetry yield high Qf which indicate the perfect crystal. And as defects and impurity degrade the perfect crystal, Qf should be reduced. As grain growth without rough microstructure is transient state to single crystal, it brings high Qf.

2)   The pseudo-tungstenbronze solid solutions with high dielectric constant of 80 to 90 have asymmetric large unit. Tilting of octahedron also affects the dielectric constant. Rattling effects in polyhedra such as TiO_6 octahedron and Al_2O_3, and SiO_4 tetrahedron.

3)   The temperature coefficient of resonant frequency might be considered on the resonation which is affected by tilting of octahedron, and volume of additional compound for reducing of TCf.

ACKNOWLEDGEMENTS

A part of this study was supported by Grant-in Aid for Scientific Research (C), and Adaptable & Seamless Technology Transfer Program (A-Step) from the Ministry of Education, Culture, Sports, Science and Technology, Japan.

REFERENCES

[1] H. Ohsato, "Science of tungstenbronze-type like $Ba_{6-3x}R_{8+2x}Ti_{18}O_{54}$ ($R=$ rare earth) microwave dielectric solid solutions", *J. Euro. Ceram. Soc.* **21**, 2703-2711 (2001).

[2] H. Ohsato, Y. Futamata, H. Sakashita, N. Araki, K.Kakimoto and S. Nishigaki, "Configuration and coordination number of cation polyhedra of tungstenbronze-type-like $Ba_{6-3x}Sm_{8+2x}Ti_{18}O_{54}$ solid solutions", *J. Eur. Ceram. Soc.* **23**, 2529-2533 (2003).

[3] M. T. Sebastian, "Dielectric materials for wireless communication", Elsevier Science Publishers, Amsterdam, 2008.

[4] H. Toraya, "Whole-Powder-Pattern Fitting without Reference to a Structural Model: Application to X-ray Powder Diffractometer Data," *J. Appl. Cryst.,* **19**, 440-447 (1986).

[5] S. Sasaki, "A Fortran Program for the Least-Squares Refinement of Crystal Structures," *XL Report, ESS,* State Univ. of New York, 1-17 (1982).

[6] F. Izumi, and T. Ikeda, "A Rietveld-analysis program RIETAN-98 and its applications to zeolites," *Mater. Sci. Forum,* **321-324**, 198-203, January 2000.

[7] B. W. Hakki, and P. D. Coleman, "A Dielectric Resonator Method of Measuring Inductive in the Millimeter Range", *IRE Trans. Microwave Theory & Tech.,* **MTT-8**, 402-410 (1960).

[8] Y. Kobayashi, and M. Katoh, "Microwave Measurement of Dielectric Properties of Low-loss Materials by the Dielectric Resonator Method", *IEEE Trans.* on **MTT-33**, 586-92 (1985).

[9] H. Ohsato, M. Imaeda, Y. Takagi, A. Komura and T. Okuda, "Microwave quality factor improved by ordering of Ba and rare-earth on the tungstenbronze-type $Ba_{6-3x}R_{8+2x}Ti_{18}O_{54}$($R=$La, Nd and Sm) Solid Solutions", Proceeding of the XIth IEEE International Symposium on Applications of Ferroelectrics, IEEE catalog number 98CH36245, pp509-512 (1998).

[10] T. Nagatomo, T. Otagiri, "M. Suzuki and H. Ohsato, "Microwave dielectric properties and crystal structure of the tungstenbronze-type like $(Ba_{1-\alpha}Sr_{\alpha})_6(Nd_{1-\beta}Y_{\beta})_8Ti_{18}O_{54}$ solid solutions", *J. Eur. Ceram. Soc.,* **26**, 1895-1898 (2006.3).

[11] Kawashima, S., Nishida, M., Ueda, I. and Ouchi, H., Dielectric properties at microwave frequencies of the ceramics in $BaOSm_2O_3TiO_2$ system, presented *at the 87th Annual Meeting, American Ceramic Society,* Cincinnati, OH, May 6, 1985 (Electronics Division Paper No.15-E-85).

[12] S. Nomura, K. Toyama and K. Kaneta, *Jpn. J. Appl. Phys.* **21**, 624-626 (1982).

[13] E. Koga, H. Moriwake, "Effects of Superlattice Ordering and Ceramic Microstructure on the Microwave Q Factor of Complex Perovskite-Type Oxide $Ba(Zn_{1/3}Ta_{2/3})O_3$", *J. Ceram. Soc. Jpn,* 767-775 (2003) (Japanese).

[14] E. Koga, H. Moriwake, K. Kakimoto and H. Ohsato, "Influence of Composition Deviation from Stoichiometric $Ba(Zn_{1/3}Ta_{2/3})O_3$ on Superlattice Ordering and Microwave Quality Factor Q", *J. Ceram. Soc. Jpn.,* 113[2], 172-178 (2005) (Japanese).

[15] E. Koga, Y. Yamagishi, H. Moriwake, K. Kakimoto and H. Ohsato, "Large Q factor variation within dense, highly ordered $Ba(Zn_{1/3}Ta_{2/3})O_3$ system ", *J. Euro. Ceram. Soc.,* **26**, 1961-1964 (2006).

[16] E. Koga, H. Mori, K. Kakimoto and H. Ohsato, "Synthesis of Disordered $Ba(Zn_{1/3}Ta_{2/3})O_3$ by Spark Plasma Sintering and Its Microwave Q Factor", *Jpn. J. Appl. Phys.,* **45(9B),** 7484-7488

(2006).

[17]E. Koga, Y. Yamagishi, H. Moriwake, K. Kakimoto and H. Ohsato, "Order-disorder transition and its effect on Microwave quality factor Q in Ba(Zn$_{1/3}$Nb$_{2/3}$)O$_3$ system", *J. Electroceram*, **17**, 375-379 (2006).

[18]H. Ohsato, F. Azough, E. Koga, I. Kagomiya, K. Kakimoto, and R. Freer, "High Symmetry Brings High Q Instead of Ordering in Ba(Zn$_{1/3}$Nb$_{2/3}$)O$_3$: A HRTEM Study", Ceramic Transactions, Volume 216, 129-136, "Advances in Multifunctional Materials and Systems" Edited by J. Akedo, H. Ohsato, and T. Shimada, Volume Editor: M. Singh, Copyright c 2010 by The American Ceramic Society, Published by John Wiley & Sons, Inc., Hoboken, New Jersey.

[19]H. Ohsato, E. Koga, I. Kagomiya and K. Kakimoto, "Origin of High Q for Microwave Complex Perovskite", *Key Eng. Mat*. 421-422 (2010) pp77-80. CSJ Seroes - Publication of the Ceramic Society of Japan - vol.19.

[20] H. Ohsato, E. Koga, I. Kagomiya and K. Kakimoto, "Phase Relationship and Microwave Dielectric Properties in the Vicinity of Ba(Zn$_{1/3}$Ta$_{2/3}$)O$_3$", *Ceram. Eng. & Sci. Proc.*, **30** (9), 25-35 (2010).

[21]M. Terada, K. Kawamura, I. Kagomiya, K. Kakimoto and H. Ohsato, "Effect of Ni substitution on the microwave dielectric properties of cordierite", *J. Eur. Ceram. Soc.*, **27**, 3045-3148 (2007.3).

[22]H. Ohsato, M. Terada, I. Kagomiya, K. kawamura, K. Kakimoto, and E-S. Kim, "Sintering Conditions of Cordierite for Microwave/Millimeterwave Dielectrics", *IEEE* **55**(5), 1082-1085 (2008).

[23]H. Ohsato, I. Kagomiya, M. Terada, and K. Kakimoto, "Origin of improvement of Q based on high symmetry accompanying Si–Al disordering in cordierite millimeter-wave ceramics", *J. Eur. Ceram. Soc.*, **30**, 315-318 (2010).

[24]H. Ohsato, J.-S. Kim, A-Y. Kim, C.-I. Cheon, and K.-W. Chae, "Millimeter-Wave Dielectric Properties of Cordierite/Indialite Glass Ceramics", *Jpn. J. Applied Physics*, **50**(9), (2011) 09NF01-1-5.

[25]K. Kugimiya, Crystallographic study on the Q of Ba(Mg$_{1/3}$Ta$_{2/3}$)O$_3$ dielectrics. Abstract for Kansai branch academic meeting held at Senri-Life Science, on the Ceramic Soc Jpn. 2003/9/5; B-20: 20; Abstract for the 10th Meeting of Microwave/Millimeterwave Dielectrics and Related Materials on the Ceram Soc. Jpn. Nagoya Institute of Technology. Japan. 2004/6/21.(Japanese)

[26]M. Andou, T. Tsunooka, Y. Higashida, H. Sugiura, and H. Ohsato, "Development of high Q forsterite ceramics for high-frequency", applications. MMA2002 Conference, 1–3 September 2002, York, UK.

[27]T. Tsunooka, M. Andou, Y. Higashida, H. Sugiura and H. Ohsato, "Effects of TiO2 on sinterability and dielectric properties of high-Q forsterite ceramics", *J. Eur. Ceram. Soc.*, **23**(14), 2573-2578, (2003).

[28]T. Tsunooka, T. Sugiyama, H. Ohsato, K. Kakimoto, M. Andou, Y. Higashida and H. Sugiura, "Development of Forsterite with High Q and Zero Temperature Coefficient t_f for Millimeterwave Dielectric Ceramics", *Key Engineering Materials*, **269**, 199-202 (2004).

[29]T. Tsunooka, H. Sugiyama, K. Kakimoto, H. Ohsato, and H. Ogawa, "Zero Temperature Coefficient tf and Sinterability of Forsterite Ceramics by Rutile Addition", J. Ceram. Soc. Jpn, Suppl., 112, S1637-S1640 (2004).

[30]H. Ohsato, T. Tsunooka, Y. Ohishi, Y. Miyauchi, M. Ando and K. Kakimoto, "Millimeter-wave dielectric ceramics with high quality factor and low dielectric constant", J. Korean Ceram. Soc.,40, 4, 350-353 (2003).

[31] J. Krupka, K. Derzakowski, M.E. Tobar, J. Hartnett, and R.G. Geyer, Meas. Sci. Tech. 10(1999).

[32] H. Ohsato, T. Tsunooka, A. Kan, Y. Ohishi, Y. Miyauchi, Y. Tohdo, T. Okawa K. Kakimoto and H. Ogawa, "Microwave-Millimeterwave Dielectric Materials" *Key Eng. Mat.*, **269**, 195-198 (2004).

[33] H. Ohsato "Research and Development of Microwave Dielectric Ceramics for Wireless Communications" *J. Ceram. Soc. Jpn.*, **113**[11], 703-711 (2005).

[34] H. Ohsato, T.Ohhashi, H. Kato, S. Nishigaki and T. Okuda, "Microwave Dielectric Properties and Structure of the $Ba_{6-3x}Sm_{8+2x}Ti_{18}O_{54}$ Solid Solutions", *Jpn. J. Appl. Phys.*, **34,** 187-191 (1995).

[35] H. Ohsato "Microwave Materials with High Q and Low Dirctric Constant for Wireless Communications", *Mater. Res. Soc. Symp.* Proc. **833**, 55-62 (2005).

[36] JIS R 1627-1996

[37] I. M. Reaney, E. L. Colla and N. Setter, "Dielectric and Structural Characteristics of Ba- and Sr-based Complex Perovskites as a Function of Tolerance Factor", *Jpn. J. Appl. Phys.* **33**, 3984-3990 (1994).

[38] H. D. Megaw, "Crystal structure of double oxides of the perovskite type", *Proc. Phys. Soc.*, **58**, 133 (1946).

[39] H. Ohsato, H. Kato, M. Mizuta, S. Nishigaki and T. Okuda, "Microwave Dielectric Properties of the Ba_{6-3x} $(Sm_{1-y}, R_y)_{8+2x}Ti_{18}O_{54}$ (R = Nd and La) Solid Solutions with Zero Temperature Coefficient of the Resonant Frequency", *Jpn. J. Appl. Phys.*, **34**, 9B, 5413-5417 (1995).

Oxide Materials for Nonvolatile Memory Technology and Applications

STABLE RESISTIVE SWITCHING CHARACTERISTICS OF Al_2O_3 LAYERS INSERTED IN HfO_2 BASED RRAM DEVICES

Chun-Yang Huang, Jheng-Hong Jieng, Tseung-Yuen Tseng
Department of Electronics Engineering and Institute of Electronics, National Chiao Tung University
Hsinchu 30010, Taiwan

ABSTRACT

Resistive switching random access memory (RRAM) has attracted extensive attention for next-generation nonvolatile memory application due to the merits of low power consumption, high speed operation, and high density integration. The resistive switching (RS) characteristics of various metal oxides have been studied. Among those materials, HfO_2 is one of the appealing materials that had received considerable attention owing to a high dielectric constant, simple composition, and its standard CMOS processes compatibility. However, the thermal stability of HfO_2 thin film is a serious issue for memory characteristics due to the low crystalline temperature. In this work, we utilized ALD growth HfO_2 thin films with inserted different amount of Al_2O_3 layers as RS layer for crystallization and RS characteristics study. From the experimental results, the crystalline temperatures depend on the amount of inserted Al_2O_3 layers. By the way, the forming voltages were modulated by using different amount of Al_2O_3 layers inserted in HfO_2 thin film, changed from 2.5 V (HfO_2) to 4.1 V (Al_2O_3). Moreover, the device shows better resistive switching performance than pure HfO_2 and pure Al_2O_3 devices, such as more stable operation voltage and higher resistive switching cycles (11000 cycles)

INTRODUCTION

Resistive random access memory (RRAM) is a promising candidate for next generation nonvolatile memory due to its simple structure, low voltage operation, high scalability, and multibit data storage.[1] It was found that transition metal oxides (TMOs) can be utilized in RRAM devices, such as ZrO_2,[2] NiO,[3] and HfO_2.[4] Among those TMOs, HfO_2 is one of the appealing materials that had considerable attention owing to its high dielectric constant (k), superior resistive switching (RS) performance, and compatible standard complementary metal oxide semiconductor (CMOS) technology process. However, the thermal stability of HfO_2 thin film is a serious issue for memory characteristics due to the low crystallization temperature (<400 °C). The RRAM devices with crystalline phase HfO_2 film suffer the RS parameters variation due to location dependent conductive filament (CF) formation. The high forming voltage is also detected in crystalline phase HfO_2 RRAM devices.[5] The high forming voltage may cause RRAM devices hard breakdown during forming process. According to previous literature, the RS behaviors are dependent on the degree of crystalline of HfO_2 film, which critically influences the device yield. Hence, this phenomenon of low crystallization temperature in HfO$_2$ is not allowed existence in further RRAM applications.

In this study, we fabricate Hf$_x$Al$_y$O films, which are the architecture with a series of complex HfO$_2$/Al$_2$O$_3$ layer by layer structure by using atomic layer deposition system (ALD), for HfO$_2$-based RRAM devices. The crystallization temperature of HfAlO thin films can be increased by increasing the number of Al$_2$O$_3$ layers in HfO$_2$ film during ALD deposition. In addition, the memory performances such as endurance, retention, and multibit storage properties are also discussed.

EXPERIMENTS

The 5-nm thin HfO$_2$, Hf$_x$Al$_y$O, and Al$_2$O$_3$ RS layers were deposited on Pt/Ti/SiO$_2$/Si substrates by using ALD at 250 °C and 0.2 Torr Ar ambient with Hf[N(C$_2$H$_5$)(CH$_3$)]$_4$, (CH$_3$)$_3$Al, and H$_2$O precursors. The Hf$_x$Al$_y$O films were the t-series of complex m-cycle HfO$_2$ layers and n-cycle Al$_2$O$_3$ layers structure, or $\{(HfO_2)_m/(Al_2O_3)_n\}_t$ multilayer architecture. The HfO$_2$ layers were deposited first and then Al$_2$O$_3$ layers during every series deposition. For example, the 5-nm Hf$_{0.7}$Al$_{0.3}$O film was composed of $t = 8$ series of mixed $m = 6$ cycle HfO$_2$ layers and $n = 1$ cycle Al$_2$O$_3$ layers. The material composition of Hf$_x$Al$_y$O films was modified by different m and n values, besides, the thickness was controlled by t value, as shown in Table I. Subsequently, the post deposition annealing (PDA) processes were carried out for X-ray diffraction (XRD) analysis at different temperatures in N$_2$ ambient for 30 s. Finally, a 50-nm thick Ti top electrode and a 20-nm thick Pt capping layer with a diameter of 150 μm were deposited by electron beam evaporation. All the electrical characteristics were performed using an Agilent 4156C semiconductor parameter analyzer.

Table I. Components of 5-nm $\{(HfO_2)_m/(Al_2O_3)_n\}_t$ multilayer architecture.

Composition	HfO$_2$	Hf$_{0.7}$Al$_{0.3}$O	Hf$_{0.55}$Al$_{0.45}$O	Hf$_{0.11}$Al$_{0.89}$O	Al$_2$O$_3$
m-cycle	1	6	3	1	0
n-cycle	0	1	1	3	1
t-series	56	8	14	14	53

RESULTS AND DISCUSSION

A typical cross-section TEM image of the Ti/Hf$_{0.7}$Al$_{0.3}$O/Pt RRAM device is shown in Fig. 1(a). The as-deposited Hf$_{0.7}$Al$_{0.3}$O film composed of complex HfO$_2$/Al$_2$O$_3$ layer by layer structure shows amorphous and non-distinguishable interface between Al$_2$O$_3$ and HfO$_2$ layers, which is indicated that the Al atoms are uniformly distributed in the Hf$_{0.7}$Al$_{0.3}$O film. Fig. 1(b) shows the XRD patterns of HfO₂ and Hf₀.₇Al₀.₃O films with as-deposited and 400 °C PDA processes, respectively. The Hf₀.₇Al₀.₃O film is still in amorphous state after 400 °C PDA process. However, the HfO₂ film shows crystalline phase. In addition, Fig. 2 reveals the crystallization temperature of Hf$_x$Al$_y$O films as a function of different Al percentage (Al % = y/[x+y]). By utilizing Al₂O₃ layers inclusion in HfO₂ film, the crystallization temperature increases from 400 °C for HfO₂ film to 1200°C for Al₂O₃ film. It can be explained that Al distributes uniformly in Hf$_x$Al$_y$O films and Al acts as a network modifier to suppress the crystallization of HfO₂ film.[6]

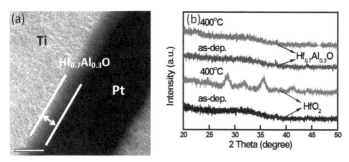

Figure 1. (a) Typical cross-section TEM image of the Ti/Hf₀.₇Al₀.₃O/Pt RRAM device. (b) XRD patterns of HfO₂ and Hf₀.₇Al₀.₃O films with as deposited and 400 °C PDA, respectively.

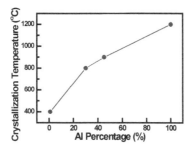

Figure 2. Crystallization temperature of different Al percentage in Hf$_x$Al$_y$O films.

The forming voltages (V$_F$) in the HfO₂ and Hf$_x$Al$_y$O RRAM devices are also dependent on Al percentage, as shown in Fig. 3. Al incorporated in the HfO₂ devices induce the increasing of the V$_F$ from 2.45 V for HfO₂ device to 4.1 V for Al₂O₃ device. On the other words, the V$_F$ can be modulated by different Al percentage inclusion in HfO₂ RRAM devices. Fig. 4 shows the comparison of V$_F$ distribution in HfO₂ device with that in Hf₀.₇Al₀.₃O device. The Hf₀.₇Al₀.₃O device reveals a much narrower V$_F$ distribution than HfO₂ device. This phenomenon is due to that the oxygen vacancies are easier to generate and assemble along Al atoms from bottom to top electrodes and the CF grows stably along Al atoms in RS layer.[7] In addition, the statistical distributions of high resistance state (HRS) and low resistance state (LRS) during resistance switching cycles are also greatly improved, as shown in Fig. 5.

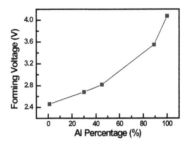

Figure 3. Forming voltage of Hf$_x$Al$_y$O devices as a function of Al percentage. The inset shows the statistical distributions of forming voltages of HfO$_2$ and Hf$_{0.7}$Al$_{0.3}$O devices.

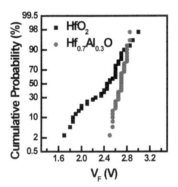

Figure 4. Statistical distributions of forming voltages of HfO$_2$ and Hf$_{0.7}$Al$_{0.3}$O devices.

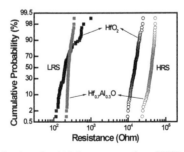

Figure 5. Resistance distributions for 100 dc sweep cycles of HfO$_2$ and Hf$_{0.7}$Al$_{0.3}$O devices. The resistances are measured at a read voltage of 0.3 V.

To further confirm the RS performance, the electrical properties of the Hf$_{0.7}$Al$_{0.3}$O device are also studied. Fig. 6 depicts the endurance characteristic of the Hf$_{0.7}$Al$_{0.3}$O device after post metal annealing (PMA) at 400°C for 30 min in vacuum ambient. The resistance ratios of HRS/LRS can be well retained after more than 11000 switching cycles under set voltage (V$_{set}$) of 0.6 V and reset voltage (V$_{reset}$) of -0.5 V applied on Ti top electrode. Fig. 7 shows the read disturbance property of the Hf$_{0.7}$Al$_{0.3}$O device under a positive voltage stress (0.3 V) at room temperature. It is clear that both HRS and LRS do not show any degradation for more than 10^4 s. In addition, Fig. 8 shows the influence of memory states before and after backing test with 200 °C in N$_2$ ambient for 1 min. The current in HRS after baking test slightly increases but the on/off ratio still remains >100 times for more than 10^3 s. From above results, the Hf$_{0.7}$Al$_{0.3}$O device exhibits more stable and uniform RS characteristics than HfO$_2$ device.

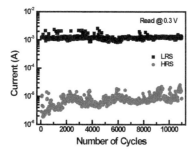

Figure 6. Endurance characteristic of Hf$_{0.7}$Al$_{0.3}$O device for 11000 switching cycles.

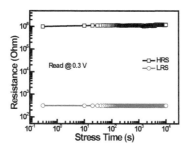

Figure 7. Read disturbance behavior for the device at room temperature.

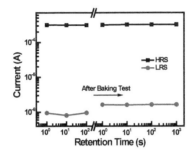

Figure 8. Retention before and after backing test. The currents are measured at a read voltage of 0.3 V.

CONCLUSION

The thermal stability can be improved by inserting Al$_2$O$_3$ layers in HfO$_2$ film. Base on experimental results, the crystallization temperature and forming voltage of HfO$_2$ based RRAM devices can be modulated by changing the number of Al$_2$O$_3$ layers in HfO$_2$ film during ALD deposition. In addition, the Hf$_{0.7}$Al$_{0.3}$O device has a similar forming voltage with HfO$_2$ device but superior thermal stability than pure HfO$_2$ device. Therefore, the resistive switching performances of Hf$_{0.7}$Al$_{0.3}$O device are investigated in this study. The device shows less variation of resistive switching operations than HfO$_2$ device. Especially, the device exhibits good memory performances, including low operation voltage, reproducible endurance, and reliable read disturbance. Above results suggest that Hf$_{0.7}$Al$_{0.3}$O device is promising for next generation nonvolatile memory application.

REFERENCES

[1]M. C. Wu, Y. W. Lin, W. Y. Jang, C. H. Lin, and T. Y. Tseng, *IEEE Electron Device Lett.*, **32**, 1026 (2011).

[2]C. Y. Lin, C. Y. Wu, C. Y. Wu, T. C. Lee, F. L. Yang, C. Hu, and T. Y. Tseng, *IEEE Electron Device Lett.*, **28**, 366 (2007).

[3]D. Ielmini, F. Nardi, C. Cagli and A. L. Lacaita, in *Proc. International Reliability Physics Symposium*, 5D.1.1 (2010).

[4]H. Y. Lee, P. S. Chen, T. Y. Wu, Y. S. Chen, C. C. Wang, P. J. Tzeng, C. H. Lin, F. Chen, C. H. Lien, and M. J. Tsai, *Tech Dig Int Electron Device Meet.*, 297 (2008).

[5]M. Lanza, G. Bersuker, M. Porti, E. Miranda, M. Nafria, and X. Aymerich, *Appl. Phys. Lett.*, **101**, 193502 (2012).

[6]W. J. Zhu, T. Tamagawa, M. Gibson, T. Furukawa, and T. P. Ma, *IEEE Electron Device Lett.*, **23**, 649 (2002).

[7]S. Yu, B. Gao, H. Dai, B. Sun, L. Liu, X. Liu, R. Han, J. Kang, and B. Yu, *Electrochem. Solid-State Lett*., **13**, H36 (2010).

IMPROVEMENT OF RESISTIVE SWITCHING PROPERTIES OF Ti/ZrO$_2$/Pt WITH EMBEDDED GERMANIUM

Chun-An Lin, Debashis Panda, Tseung-Yuen Tseng[a]

[a]Tel.: (+886)-3-5731879. FAX: (+886)-3-5724361. Electronic mail: tseng@cc.nctu.edu.tw.

Department of Electronics Engineering and Institute of Electronics, National Chiao Tung University,

Hsinchu 300, Taiwan

ABSTRACT

In this study, we construct the Ti/ZrO$_2$/Ge(5nm)/ZrO$_2$/Pt structures with various positions of embedded Ge in ZrO$_2$ films. After depositing a Ge layer, a 600 °C rapid thermal annealing is carried out. Compared to other Ge positions, the lowest forming voltage and the most stable resistive behavior are observed in the cell with a Ge layer near the top electrode. The curve fitting of high resistance state and low resistance state shows that Schottky emission in reset process and the ionic conduction during set process. The improved switching properties could be related to the formation of Germanium oxide and the defect concentration reduction after annealing process.

INTRODUCTION

Resistive switching random access memory (RRAM) is one of potential next-generation memories due to its high scalability, low power consumption, high switching speed and high switching cycles. RRAM switching modes could be divided by three types: bipolar, unipolar, and nonpolar mode. The switching mechanism of bipolar ZrO$_2$-based RRAM devices is generally explained by growth and rupture of conducting filaments consisting of oxygen vacancies, which correspond to the low resistance state (LRS) and the high resistance state (HRS) of the memory, respectively.

To improve switching properties, the embedding metal in oxide layer has been widely investigated, such as ZrO$_2$/Au/ZrO$_2$,[1] ZrO$_2$/Mo/ZrO$_2$,[2] TiO$_2$/Pt/TiO$_2$,[3] ZrO$_2$/Co/ZrO$_2$,[4] and so on. However, the reports about inserting a semiconductor layer into resistive switching memories are quite limited. This paper reports that the embedding Germanium layer in RRAM cell can effectively reduce the forming voltage, decrease the limited current, and increase switching cycles.

DEVICE FABRICATION

A ZrO$_2$ thin film was fabricated by a sputter on the Pt/Ti/SiO$_2$/Si substrate. Next, a 5 nm Ge layer was grown on the ZrO$_2$ thin film by electron beam evaporation, followed by 600 °C rapid thermal annealing for 30 s. Then, a ZrO$_2$ layer was sputtered on the Ge film. Finally, a 50 nm Ti top electrode was deposited by electron beam evaporation. By changing the thicknesses of the first and second ZrO$_2$ films in the present devices, the same bottom and top electrodes with three different embedded Ge positions was fabricated. The thickness of 20/5/3 of ZrO$_2$/Ge/ZrO$_2$ is literally written as EB (Ge near Bottom electrode), 12/5/12 as EM and 3/5/20 as ET, separately. Besides, the Ti\28 nm ZrO$_2$\Pt devices with and without annealing were fabricated for comparison.

THE FORMING AND THE SET/RESET VOLTAGE COMPARISON

The distributions of the forming voltage based on 10 random samples of each structure are shown in Figure 1(a). The ET structure performs the most uniform and the lowest forming voltage than other structures, where most are between 2 and 4 V. It also demonstrates that the structures with the embedded a Ge layer have the lower forming voltage. Figures 1(b) and (c) show the distributions of reset/set voltage, only the EB structure reveals non-uniform distributions of voltage, while others are uniform. The EM structure reveals the distribution of reset voltage between -1 and -1.5 V, and set voltage is between 0.6 and 1.3 V. For ET, reset voltage is mainly distributed between -0.8 and -1.5 V, and set voltages are almost between 0.7 and 1.5 V.

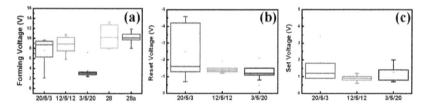

Figure 1. (a) Forming voltage distributions of the EB, EM, ET structures, control samples with and w/o annealing, (b) Reset voltage distributions of the EB, EM, and ET structures. (c) Set voltage distributions of the EB, EM, and ET structures.

Figures 2(a)-(c) show the 10-time dc cycling I-V curves of the EB, EM, and ET structures, respectively. All the three structures are bipolar switching. The limited currents of the EB, EM and ET structures are 2, 3, and 0.3 mA, respectively. Figures 2(d)-(e) demonstrate the LRS and HRS currents of the EB, EM, and ET structures under 100-cycles, respectively. Figures 2(d) and (e) show unstable on state, which may be due to different filament growth paths. However, the distributions of resistance states are stable in on state, while the off state is unstable, as shown in Figure 2(f). Hence, the ET structure is advantageous for ZrO$_2$-based RRAM devices in the consideration of electrical properties.

SWITCHING PERFORMANCE AND RELIABILITY TEST

Figure 3(a) shows the TEM image of the ET structure which has better electrical performance than the EB and EM structure. Figure 3(b) demonstrates the switching cycle of the ET structure is up to 8000, and the on/off ratio is 5. The plot also shows that the large distribution of HRS lowers the resistance ratio.

The on state and off state of the ET structure device keep stable under an bias voltage of 0.3 V after 10^4 s, as shown in Figure 4(a). The retention test under 85 °C is depicted in Figure 4(b) which shows the ET structure remain steady after 10^3 s, however, the HRS current becomes higher after 3×10^3 s, which leads to a readout error.

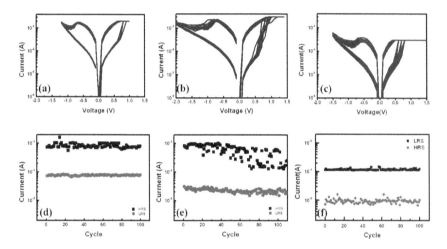

Figure 2. (a)-(c) Typical I-V curves of the EB, EM, and ET structures after 10 cycles, respectively. (d)-(f) HRS and LRS currents of the EB, EM, and ET structures after 100 cycles, respectively.

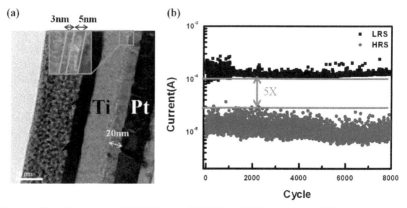

Figure 3. The ET structure (a) TEM image (b) HRS and LRS currents of DC endurance test.

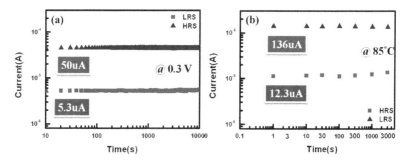

Figure 4. The ET structure (a) HRS and LRS current of stress test under biasing 0.3 V for 10^4 s (b) HRS and LRS currents of retention test at 85℃ for 3×10^3 s.

CURVE FITTING DURING SET/RESET PROCESS

It's important to understand the conduction mechanism of the HRS and LRS for exploring the RS mechanism of Ge embedded ZrO₂ device. Schottky emission fitting for the HRS is satisfied. Schottky emission can be expressed as[5]:

$$ J = A * T^2 \exp\left[\frac{-q(\emptyset_B - \sqrt{qE/4\pi\varepsilon_i})}{kT}\right] \tag{1} $$

where J is the current density, T is the absolute temperature, ε_i is the dynamic dielectric constant, q is the elementary charge, φ_B is the barrier height, and k is the Boltzmann constant.

Figure 5(a) presents the plot of $Ln(J/T^2)$ as a function of $V^{1/2}$ of the HRS currents of the EB,

EM, and ET structures, respectively, indicating that the conduction of carrier in the thin film is Schottky emission. In Figure 5(b), the temperature dependence of the ET structure under biasing -0.4 V also obeys Schottky emission in the temperature ranging from 298 K to 423 K.

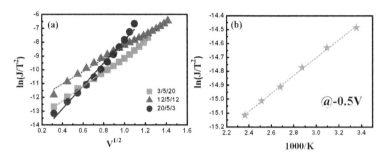

Figure 5. (a) The curve fitting of Schottky emission for the HRS currents. (b) The temperature dependence of the HRS current at −0.5 V for the ET structure.

The fittings of the LRS current of the ZrO_2 based RRAM devices with different embedded Ge positions are also performed. The ionic conduction is expressed as[5]:

$$J \sim \frac{V}{T} e^{-\frac{d}{T}} \tag{2}$$

where J is the current density, T is the absolute temperature , and d is the thickness of the $ZrO_2/Ge/ZrO_2$ film.

Figure 6(a) shows the plot of Ln(J) as a function of Ln(V) of the HRS currents of the devices. To further verify the mechanism, the temperature dependence of the ET structure at 0.5 V on the plot $Ln(J/T^2)$ versus (1000/K) is demonstrated in Figure 6(b). Both curve fittings imply that the Ge ion conducting plays an important role during set process.

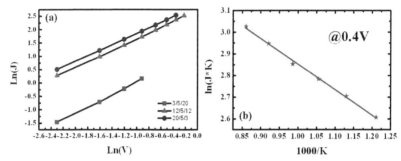

Figure 6. (a) The curve fittings of ionic conduction for LRS current of the EB, EM, and ET structures. (b) The temperature dependence of the LRS current at 0.4 V for the ET structure.

CONCLUSION

The embedded Ge layer near top electrode in Ti/ZrO₂/Pt effectively reduces the forming voltage. Low compliance current (300 μA), low off state current (5 μA) and higher DC cycles are also achieved in the ET structure, which is owing to the conduction of Ge ion with the assistance of on-state ionic-emission fitting.

REFERENCES

[1]W. Guan, S. Long, R. Jia, and M. Liu, Nonvolatile resistive switching memory utilizing gold nanocrystals embedded in zirconium oxide, *Appl. Phys. Lett.*, **91**, 062111 (2007).

[2]S.-Y. Wang, D.-Y. Lee, T.-Y. Huang, J.-W. Wu and T.-Y. Tseng, Controllable oxygen vacancies to enhance resistive switching performance in a ZrO₂-based RRAM with embedded Mo layer, *Nanotech.*, **21**, 495201(2010).

[3]W.-Y. Chang, K.-J. Cheng, J.-M. Tsai, H.-J. Chen, F. Chen et al., Improvement of resistive switching characteristics in TiO₂ thin films with embedded Pt nanocrystals, *Appl. Phys. Lett.*, **95**, 042104 (2009).

[4]M.-C. Wu, T.-H. Wu, and T.-Y. Tseng., Robust unipolar resistive switching of Co nano-dots embedded ZrO₂ thin film memories and their switching mechanism, *J. Appl. Phys.*,**111**, 014505 (2012).

[5]S. M. Sze, *Physics of Semiconductor Devices*, 2nd ed. New York:Wiley, 1981.

NONVOLATILE MEMORIES USING SINGLE ELECTRON TUNNELING EFFECTS IN SI QUANTUM DOTS INSIDE TUNNEL SILICON OXIDE

Ryuji Ohba

Advanced Memory Device Technology Laboratory, Center for Semiconductor Research & Development, TOSHIBA Corporation

800, Yamanoisshiki-cho, Yokkaichi-shi, Mie-ken, 512-8550, JAPAN

E-mail / Tel : ryuji.ooba@toshiba.co.jp / +81-59-390-7455

ABSTRACT

For further memory device scaling in NAND Flash, we investigate new tunneling effects (double junction tunnel) in conventional field-effect-transistor (FET) memories that have charge storage part between channel and control gate. Instead of the silicon oxide tunnel layer between charge storage part and channel, we use double tunnel junction structure, in which Si nanocrystal (Si quantum dot) layer is lying between 2 thin tunnel silicon oxides. As a result of quantum effects in the Si nanocrystals, the memory device characteristics can be improved remarkably, and smaller Si nanocrystal size can lead to better memory characteristics. It is noted that the most significant memory scaling problems (high voltage operations and thick insulators) can be resolved, since Si nanocrystal layer double junction can simultaneously attain low-voltage high-speed w/e, thin tunnel layer thickness and non-volatility, We also show that sub-10nm device scaling, which is close to the physical limit of scaling, is possible by sub-1nm Si quantum dot size control inside tunnel layer. Thus, small Si dot size control inside tunnel layer is the key issue. It is concluded that Si nanocrystal layer double junction memory realized by further Si dot scaling is a very promising candidate for future memory devices.

SINGLE ELECTRON TUNNELING

In conventional nonvolatile flash memory cells, information charge is written or erased by applying high write / erase voltages using usual FN tunneling phenomena through a thick tunnel insulator (about 8nm-thick silicon oxide $SiO2$), and a long charge retention time is possible owing to the thickness of the tunnel oxide. Recently, low-voltage operation and thin tunnel oxide have been strongly demanded for future flash memory cell size scaling.

So as to improve tunnel oxide characteristics, we introduce single-electron tunneling phenomena [1, 2] into the flash memory tunnel insulator. When an electron tunnels through a very small nm-scale conductive structure, the tunnel characteristics change greatly because each electron energy state varies largely due to the nm-scale structure. One of the most important origins of the change in tunnel characteristics is Coulomb blockade.

The tunnel current suppression due to single electron charging energy is called Coulomb blockade [1, 2]. When a nm-scale conductive island (dot) has a total capacitance C, the electrostatic charging energy $q^2/2C$ is necessary for one electron existence in the island, where q is the elementary charge. When the nm-scale island is very small, C will be very small. Then, the charging energy $q^2/2C$ can be very large. We can design single-electron tunneling using the large charging energy. In the single-electron tunnel design, the most typical structure is the double tunnel junction.

The double tunnel junction [3, 4] is the direct connection of 2 tunnel junctions separated by a nm-scale conductive island. In order to attain a single-electron tunnel through the double tunnel junction, the charging energy $\Delta E_{CB} = q^2/2C$ has to be supplied for each single-electron tunnel. We can use the charging energy ΔE_{CB} for charge retention improvement. The most typical structure is the single electron memory.

A typical single electron memory is a direct connection of a double tunnel junction and a capacitance [5, 6, 7]. A memory hysteresis effects due to ΔE_{CB} can appear in the single electron memory based on the double junction, while only a single tunnel junction resistance can induce a memory hysteresis in the single electron box which is a direct connection of a single tunnel junction and a capacitance [8]. We notice that, if we can use the double tunnel junction in tunnel layers in nonvolatile flash memory cells, the memory performance can be improved by ΔE_{CB} due to nm-scale islands within tunnel layers.

SI QUANTOM DOTS INSIDE TUNNEL SI OXIDE

We have proposed a novel Si nanocrystal (Si dot) nonvolatile memory cells that has double tunnel junctions in the tunnel insulators [9, 10, 11, 12]. In these memory cell structure, tunnel insulators consist of two thin tunnel Si oxide layers separated by Si nanocrystals, and information charge storage parts exist immediately on the double junction tunnel layers, and there is a block insulators between the charge storage parts and the control gates. Since the information electrons tunnel through double tunnel junction, a memory hysteresis effects are enhanced by single-electron energy variation as in the single electron memory.

In the double junction tunnel memory cells, a long retention time will be possible, since the charge leak between the charge storage part and the channel can be suppressed by an energy barrier due to quantum confinement and Coulomb blockade in the Si nanocrystal. If the nm-scale island is a semiconductor as Si nanocrystal, the quantum confinement energy, which comes from crystal kinetic energy, plays the same important role as the Coulomb blockade energy that comes from electrostatic energy, because of the lower density of states in a semiconductor dot than that in a metal dot. Since the energy barrier due to quantum confinement and Coulomb blockade increases with Si dot downscaling, the charge retention will be improved further by making lower Si dot size smaller. We show that charge retention can be improved exponentially by Si dot size downscaling [10].

It is shown that the exponential retention improvement due to the Si dot downscaling is attained, keeping almost the same write / erase speed [10, 11]. The high-speed write / erase is achieved, because the leak suppression due to the single electron energy barrier in the Si dot is effective only in a low voltage region within a blockade voltage. These results show that the double junction tunnel layer memory cells are promising for future low-voltage non-volatile memory.

We also clarify a characteristic effect in double junction tunnel using quantum dot, tunnel penetration disappearance, which is extremely advantageous for nonvolatile memory and never occurs in other band-engineered multilayer tunnel dielectrics structures [13, 14]. Therefore, the double tunnel junction using Si nanocrystalline layer is very promising for future memory.

SUB-10NM MEMORY DEVICE

We experimentally demonstrate a short-gate-length SONOS-type memory device where the charge storage part is silicon nitride. By using very small (close to 1nm) Si quantum dot inside tunnel layers, 15nm-gate-length bulk-planar SONOS-type memory device, which has Si nanocrystalline layer lying between double tunnel oxides, retains 2.7 decades memory window for 10 years below 10 V write / erase (w/e) voltages. Sub-threshold slope (S-factor) is controlled by source/drain (S/D) junction depth and channel concentration [15]. Experimental evidence of a remarkable advantage in trade-off between charge retention and w/e speed is shown clearly [16], and it is shown that further device scaling and improvement are possible by Si nanocrystal size downscaling. These results show that double tunnel junction SONOS-type memory is a strong candidate for future memory owing to the single-electron tunneling phenomena.

We have shown the possibility of memory device scaling down to sub-15nm region. However, infinitesimal scaling of memory device size is principally impossible from a physical viewpoint. We discuss a physical limit of FET-type memory device scaling due to source-to-drain direct tunneling [17, 18], which will be remarkable in sub-10nm region.

10nm-gate-length bulk-planar SONOS-type memory device, where 1nm Si nanocrystals are lying between 1nm-thick double tunnel oxides, retains 2.6 decades memory window for 10 years at less than 13 V w/e voltages. Moreover, 8nm-gate-length double junction SONOS device can show the same excellent characteristics by realizing source / drain (S/D) direct tunnel sub-threshold [19]. These results show that further device scaling down to sub-10nm region is possible utilizing S/D direct tunnel sub-threshold. All the short gate length memory devices show excellent characteristics and the possibility of further improvement by Si dot downscaling. Double junction SONOS is a promising candidate for sub-10nm region, within the physical limit of S/D direct tunneling.

CONCLUSION

Si quantum dot double junction tunnel layers in FET-type nonvolatile memories have been studied. We have shown that charge retention is improved greatly keeping a high-speed w/e, compared to the usual single-layer tunnel Si oxide, due to Coulomb blockade and quantum confinement in nm-scale Si quantum dots. By making Si dot size smaller, the charge retention can be further improved exponentially, keeping almost the same w/e speed. We have also shown a remarkable advantage for nonvolatile memory, tunnel penetration disappearance, which is characteristic of quantum dot double junction. This effect also shows a clear principal advantage of double tunnel junction in comparison with high-k multilayer tunnel structures.

By forming small (about 1nm) Si nanocrystal dots between the tunnel junctions, 10nm double junction SONOS devices shows an excellent non-volatility at low (about 10V) w/e voltages. Sub-10nm scaling has also been studied experimentally, and we show that 8nm double junction SONOS device will show excellent performance comparable to that in longer-gate device if we realize S/D direct tunnel S-factor, which is the physical limit of S factor. Notice that double junction tunnel layer enables sub-10nm device scaling close to the physical limit.

The quantum dot double junction tunnel layer can attain a low-voltage high-speed w/e, thin tunnel layer thickness and non-volatility simultaneously owing to the single-electron tunneling phenomena. We conclude the double tunnel junction tunnel layer memory cell is promising for future nonvolatile memory close to the physical limit in sub-10nm scaling. The control of small Si quantum dot formation will be the key issue.

REFERENCES

[1] D.V Averin and K. K. Likharev : *Mesoscopic phenomena in solids* , p.173 ,chapter 6 (Elsevier Science , 1991. Editors, B.L. Altshuler, P.A. Lee, R.A. Webb)

[2] H.Grabert and M.H.Devoret (Editor) : *Single charge tunneling Coulomb Blockade phenomena in Nanostructures*, (Plenum Press) (1992).

[3] Grabert, H., Ingold, G.-L., Devoret, M.H., Esteve, D., Pothier, H., Urbina, C., "Single electron tunneling rates in multijunction circuits", Zeitschrift fur Physik B Condensed Matter 84 (1), pp. 143-155 (1991)

[4] Higurashi, H., Iwabuchi, S., Nagaoka, Y., "Coulomb blockade and current-voltage characteristics of ultrasmall double tunnel junctions with external circuits", Physical Review B 51 (4), pp. 2387-2398 (1995)

[5] D. V. Averin and K. K. Likharev, *Single charge tunneling Coulomb Blockade phenomena in Nanostructures*, Chapter 9, p.311 (Plenum Press, edited by H.Grabert and M.H.Devoret) (1992).

[6] N. Nakazato, R. J. Blaikie, J. R. A. Cleaver and H. Ahmed, "Single-electron memory", Electronics Lett. 29 no.4 pp.384-385 (1993)

[7] Stone, N.J. and Ahmed, H, "Silicon single electron memory cell", Applied Physics Letters 73 (15), pp. 2134-2136 (1998)

[8] Lafarge, P., Pothier, H., Williams, E.R., Esteve, D., Urbina, C., Devoret, M.H., "Direct observation of macroscopic charge quantization", Zeitschrift fur Physik B Condensed Matter 85 (3), pp. 327-332 (1991)

[9] R.Ohba, N.Sugiyama, K.Uchida, J.Koga and A.Toriumi, "Nonvolatile Si Quantum Memory with Self-Aligned Doubly-Stacked Dots", Int. Electron Device Meet. Tech. Dig. (IEEE 2000) p. 313-316.

[10] R. Ohba, N. Sugiyama, K. Uchida, J. Koga and A. Toriumi, "Nonvolatile Si Quantum Memory with Self-Aligned Doubly-Stacked Dots", IEEE Trans. on Electron Devices vol.49, pp1392-1398, 2002

[11] R. Ohba, N. Sugiyama, J. Koga and S. Fujita, "Silicon Nitride Trap Memory with Double Tunnel Junction", Digest of Int. Symp. on VLSI Technology , 2003, pp.35-36.

[12] R. Ohba Y. Mitani, N. Sugiyama and S. Fujita, "Impact of Stoichiometry Control in Double Junction Memory on Future Scaling", Int. Electron Device Meet. Tech. Dig., p.897-900 (2004)

[13] R. Ohba, Y. Mitani, N. Sugiyama, M. Matsumoto and S. Fujita, " Double Junction Tunnel using Si Nanocrystalline Layer for Nonvolatile Memory Devices", Jpn. J. Appl. Phys. 50 no.4 (2011)

[14]Tseung-Yuen Tseng and Simon M. Sze (Editor), *NONVOLATILE MEMORIES, Materials, Devices and Applications*, vol.1, ch.8, "Nonvolatile Memories Using Si Quantum Dot Double-Junction Tunnel Layers"

[15] R. Ohba Y. Mitani, N. Sugiyama and S. Fujita, "15 nm Planar Bulk SONOS-type Memory with Double Junction Tunnel Layers using Sub-threshold Slope Control", Int. Electron Device Meet. Tech. Dig., p. 75 - 78 (2007)

[16] R. Ohba Y. Mitani, N. Sugiyama and S. Fujita, "25 nm Planar Bulk SONOS-type Memory with Double Tunnel Junction", Int. Electron Device Meet. Tech. Dig. p. 959 - 962 (2006)

[17] J. Wang and M. Lundstrom, "Does source-to-drain tunneling limit the ultimate scaling of MOSFETs?",Int. Electron Device Meet. Tech. Dig., pp. 707 - 710, (2002).

[18] H. Wakabayashi, T. Ezaki, M. Hane, T. Ikezawa, T. Sakamoto, H. Kawaura, S. Yamagami, N. Ikarashi, K. Takeuchi, T. Yamamoto, and T. Mogami, "Transport properties of sub-10-nm planar-bulk-CMOS devices", Int. Electron Device Meet. Tech. Dig., pp. 429 - 432, (2004)

[19] R. Ohba Y. Mitani, N. Sugiyama and S. Fujita., "10 nm Bulk-Planar SONOS-type Memory with Double Tunnel Junction and Sub-10 nm Scaling Utilizing Source to Drain Direct Tunnel Sub-threshold", Int. Electron Device Meet. Tech. Dig. pp. 894 - 897 (2008)

Resistive Switching and Rectification Characteristics with CoO/ZrO$_2$ Double Layers

Tsung-Ling Tsai, Jia-Woei Wu, and Tseng-Yuen Tseng
Department of Electronics Engineering and Institute of Electronics, National Chiao Tung University, Hsinchu 30010, Taiwan

ABSTRACT

We have observed the unipolar resistive switching with double layers of sputtered CoO/ZrO$_2$. The thicknesses of CoO/ZrO$_2$ were 20 and 30 nm deposited by sputtering, respectively. Because of the different oxygen concentration, it formed larger size filament in CoO layer than that in ZrO$_2$ layer after forming process. During OFF process, a large amount of current flows through the conducting filament, the narrower filament in the ZrO$_2$ layerwould be ruptured because of the locally Joule heating, then we observed the unipolar behavior in this structure. About 400 cycling times of unipolar switching characteristic were obtained. However, due to the p-type conductivity for CoO and the n-type conductivity for ZrO$_2$, it revealed the rectified behavior when we change to the proper thickness of the films and the ratio of sputtering gases(Ar,O$_2$). It has the potential application of 1D1R structure in the future.

INTRODUCTION

Nowadays, resistive random access memory (RRAM) has a lot potential of next generation memory because of its high density integration, high operation speed, low power consumption[1]. In addition to ZrO$_2$[2], there is a lot variety of metal oxides materials to be used in RRAM, such as HfO$_2$[3] and NiO[4]. It stores the data by the different resistance value(a low resistive ON state and a high resistive OFF state). By applying the appropriate voltage, the memory cell can be switched into two different states continuously.

The detailed resistive switching mechanisms are still unclear, where material properties and fabrication processes have considerably influence on them. However, the most acceptable and popular mechanism was believed to be the formation/rupture of conducting filament with oxygen ions/vacancies migration within RRAM device according to the reports from various research groups[3~5]. For the bipolar switching mode, the switching behavior is considered to the migration of oxygen ions driven by electric field. However, for the unipolar switching, the filament is ruptured by the Joule heating effect from the large amount of current flow through the filament.

But for the application, it is necessary to construct the memory cell in array structure. However, misreading might happened in array structure when the current flow through the other low resistance memory cells with selected cell in high resistance state. In order to solve the problem, there are several structure with selecting element such as transistor, BJT, diode are reported. Among these different type of devices, 1diode with 1RRAM(1D1R) take a large advantages than the others due to its smaller cell size, simpler design and low temperature in fabrication.

In this work, the p-CoO/n-ZrO$_2$ double layers with unipolar switching behavior are reported. Furthermore, the rectified characteristics are also demonstrated by using the same materials p-CoO/n-ZrO$_2$ with different fabricated condition.

UNIPOLAR SWITCHING BEHAVIOR OF DOUBLE LAYERS DEVICE

A CoO/ZrO$_2$ device was deposited by sputtering with the same gas ratio Ar:O$_2$=1:2. Both the top and bottom electrodes are Pt fabricated by E-beam evaporation. The thickness of Pt/CoO/ZrO$_2$/Pt structure is shown by cross-sectional transmission electron microscopy (TEM) image in Fig. 1. In Figs. 2(a) and (b), the device shows the reversible unipolar switching behavior and its endurance with 400 cycles. After forming process (V$_f$=5V), the device is switched from original state to the low resistance state. The filaments in ZrO$_2$ layer are narrower than that in CoO because of the oxygen concentration in ZrO$_2$ layer is higher than in CoO layer which could be confirmed by X-ray photoelectron spectroscopy(XPS) analysis in Fig. 3 . During the reset process, by applying to the 1 V(V$_{off}$), a large amount of current flow through the filament, the narrower filament in ZrO$_2$ layer would be ruptured due to locally Joule heating effect and switch the device to high resistance state[6].

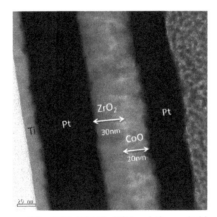

Figure 1.TEM image of the Pt/CoO/ZrO$_2$ /TiN resistive switching memory device.

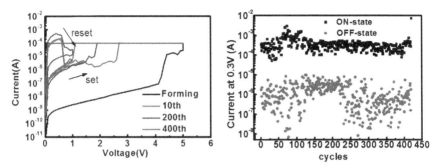

Figure 2.(a) Typical unipolar RS I-V curve of the Pt//CoO/ZrO$_2$/Pt device.(b)Endurance of the resistive switching

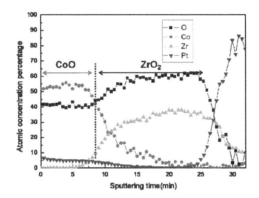

Figure 3.XPS composition depth profile of the Pt//CoO/ZrO$_2$/ Pt memory stack.

RECTIFIED PROPERTIES OF OXIDE DIODE

We choose the oxide diode that is because it can be fabricated at the room temperature, then the oxide diodes are suitable for the integration and 3D application. Due to the properties of metal deficiency and oxygen deficiency, CoO and ZrO$_2$ presents p-type and n-type behavior, respectively. Fig. 4 demonstrates the rectified I-V curve with large current at forward bias and low current at reverse bias condition. A CoO/ZrO$_2$ device was sputtered with the gas ratio of Ar:O$_2$=1:8 and Ar:O$_2$=12:6, respectively. It shows the better performance with large rectified ratio (5×10^3) and larger current density than other conditions due to more carrier concentration and smaller series resistance.

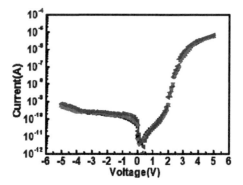

Figure 4.I-V curves of the Pt/CoO/ZrO₂/Pt oxide diode

CONCLUSION

We have demonstrated that the Pt/CoO/ZrO₂/Pt device shows the unipolar switching behavior due to the different size of filaments in CoO and ZrO₂ layer. Furthermore, we have constructed the oxide diode with the same structure Pt/CoO/ZrO₂/Pt but with different sputtered condition. The Pt/CoO/ZrO₂/Pt structure takes a large potential for the 1D1R application from it resistive switching and rectified properties.

REFERENCES

1. T.Y. Tseng and S.M. Sze, "Introduction" in Chapter 1 of Nonvolatile Memories: Materials, Devices, and Applications, Vol.1, edited by T.Y. Tseng and S.M. Sze, 2012

2. H. Kim, P. McIntyre, C. O. Chui, K. Saraswat and S. Stemmer, "Engineering chemically abrupt high-k metal oxide/silicon interfaces using an oxygen-gettering metal overlayer," J. Appl. Phys., **96**, 3467 (2004).

3. C. Y. Lin, C. Y. Wu, C. Y. Wu, and T. Y. Tseng, and C. Hu, "Modified resistive switching behavior of ZrO₂ memory films based on the interface layer formed by using Ti top electrode," J. Appl. Phys., **102**, 094101 (2007).

4. U. Russo, D. Ielmini, C. Cagli, A. L. Lacaita, S. Spiga, C. Wiemer, M. Perego and M. Fanciulli, "Conductive-filament switching analysis and self-accelerated thermal dissolution model for reset in NiO-based RRAM," Tech. Dig. - Int. Electron Devices Meet. 2007, 775.

5. W. Shen, R. Dittmann, and R. Waser, "Reversible alternation between bipolar and unipolar resistive switching in polycrystalline barium strontium titanate thin films," J. Appl. Phys., **107**, 094506 (2010).

6. M. H. Lin, M. C. Wu, C. Y. Huang, C. H. Lin, and T. Y. Tseng, "High-speed and localized

resistive switching characteristics of double-layer SrZrO3 memory devices," J. Phys. D: Appl. Phys., **43**, 295404 (2010).

RESEARCH OF NANO-SCALED TRANSITION METAL OXIDE RESISTIVE NON-VOLATILE MEMORY (R-RAM)

ChiaHua Ho, Cho-Lun Hsu, Chun-Chi Chen, Ming-Taou Lee, Hsin-Hau Huang[$%], Kai-Shin Li, Lu-Mei Lu, Tung-Yen Lai, Wen-Cheng Chiu, Bo-Wei Wu, MeiYi Li, Min-Cheng Chen, Cheng-San Wu, Yi-Ping Hsieh[#], and Fu-Liang Yang*

National Nano Device Laboratories (NDL) / National Applied Research Laboratories (NARL), Hsinchu City, Taiwan

[$] National Chio-Tung University, Hsinchu City, Taiwan
[%] Mesoscope Technology Co. Ltd., Taipei, Taiwan
[#] Institute of Edu. & Center of Teacher Edu., National Cheng Kung University, Tainan, Taiwan
[*] Corresponding to: flyang@ndl.narl.org.tw; Tel: (886) 357-26100 ext. 7701

ABSTRACT

We functionally demonstrated both Tungsten-Oxide (WO_x) and Hafnium-Oxide (HfO_x) based Transition Metal Oxide Resistive Random-Access Memory (TMO R-RAM) devices with sub-10nm scaling by Nano Injection Lithography technique. Instead of similar results of large R-RAM devices, nano-scaled WO_x and HfO_x devices beyond 10nm were found to exhibit obvious differences on programming schemes, operation window, and reliabilities. Heat dissipation dynamics plays a dominant role of R-RAM storage mechanism beyond 10nm.

INTRODUCTION

Being continuously scalable beyond current 1X nm node NAND-type floating-gate and charge-trapping (CTF) based Flash memories, the transition metal oxide based Resistive Random Access Memories (R-RAM) are much attractively studied for candidate of next generation non-volatile emerging semiconductor memories [1-2]. In recent years, the functional demonstrations of the transition metal oxide based R-RAM with 9~10nm scaled dimension were successfully achieved [3-5], indicating highly scalable characteristics of transition metal oxide based R-RAM. The scaling limitation of transition metal oxide R-RAM becomes the highly investigated focusing in advanced non-volatile memory field.

In this paper, the 9×9 nm active area device studies of Tungsten-Oxide (WO_x) and Hafnium-Oxide (HfO_x) based transition metal oxide R-RAMs are undertaken. Different programming characteristics by heat dissipation dynamics in conductive filaments are proposed to model transition metal oxide R-RAM storage mechanism beyond 10nm.

EXPERIMENTS AND RESULTS

Well-discussed semiconductor fabrication fully compatible transition metal oxide thin films, thermally oxidized non-stoichiometric WO_x [6] and atomic layer deposition (ALD) deposited HfO_x [4], are utilized to simultaneously study R-RAM device beyond 10nm by Nano Injection Lithography (NIL) technique with Platinum- and Carbon-hardmasks by reaction of $(CH_3)_3CH_3C_5H_4Pt$ and Naphthalene ($C_{10}H_8$) precursors [7], respectively. Figure 1 shows the CD-SEM images of nano injection lithography prepared Platinum array mask with 18nm X- and Y-pitch. Word -line (WL) and bit-line (BL) from the array to probe pads connect individual cells for electrical testing. Top-electrodes neighboring the selected cells are covered with dielectric SiO_2 by Nano Injection Lithography. Interconnect BL are also defined by Nano Injection Lithography.

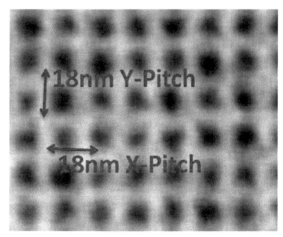

Figure 1. CD-SEM images of nano injection lithography (NIL) prepared Pt array mask with 18nm X- and Y-pitch.

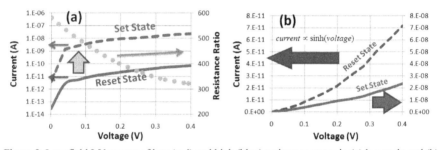

Figure 2. Low field I-V curves of low (red) and high (blue) resistance states in (a) log-scale and (b) linear-scale. Green dot-curve is the resistance ratio. Both states can be well fitted by voltage-dependent hopping transport, equation in (b), with different barrier heights.

The low voltage I-V curves of the Reset state and the Set state, shown in Figure 2, can be well fitted by hopping transport model [11].

Contrary to similar results of large R-RAM devices, the thermally oxidized non-stoichiometric WO_x and atomic layer deposition deposited HfO_x devices beyond 10nm were found to exhibit obvious differences on both thermal stability and reversible operation window:

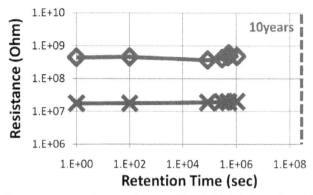

Figure 3. Room-temperature data retention test of 9nm half-pitch thermally oxidized non-stoichiometric WO_x memory cell programmed to different states.

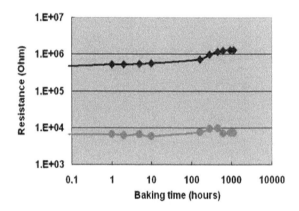

Figure 4. 250°C thermal stability testing of atomic layer deposition deposited HfO_x based R-RAM device with sub-10nm cell size.

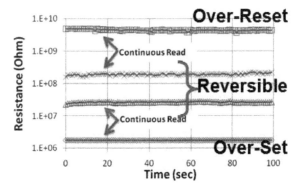

Figure 5. Optimized room-temperature operated multi-level cell (MLC) capability of 9nm thermally oxidized non-stoichiometric WO_x based R-RAM device. The intermediate states are reversible, indicating 1 order of magnitude resistance operation window.

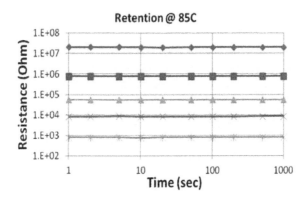

Figure 6. Optimized 85°C operated multi-level cell (MLC) capability of sub-10nm atomic layer deposition deposited HfO_x based R-RAM device, indicating at least 2 orders of magnitude resistance operation window.

Thermal Stability

Figure 3 and Figure 4 respectively present high resistance state (HRS) and low resistance state (LRS) thermal stability of 9nm thermally oxidized non-stoichiometric WO_x and sub-10nm atomic layer deposition deposited HfO_x memory cells. The former and later are respectively tested at room-temperature and 250°C baking environments, indicating atomic layer deposition deposited HfO_x based transition metal oxide R-RAM cell beyond 10nm still exhibits comparable thermal stability to current 1X nm technology node NAND-type Flash memory. The thermally oxidized non-stoichiometric WO_x based transition metal oxide R-RAM, however, behaves serious temperature dependent HRS/LRS window degradation while scaling down to beyond 10nm.

Reversible Operation Window

Figure 5 and Figure 6 respectively reveal the optimized operated multi-level cell (MLC) capability of 9nm thermally oxidized non-stoichiometric WO_x and sub-10nm atomic layer deposition deposited HfO_x based transition metal oxide R-RAM memory cells. The intermediate states are reversible by controlling bipolar operation with 1 μsec pulse with different current level, indicating atomic layer deposition deposited HfO_x based transition metal oxide R-RAM were found to exhibit better reversible resistance switching window (> 100X) than thermally oxidized non-stoichiometric WO_x based one (~ 10X) in $I_{prog} < 1$ μA range.

Possible Mechanism

The memory scale of transition metal oxide R-RAM beyond 10nm is approached to the conduction filament size. This indicates the boundary conditions of transition metal oxide R-RAM would affect the switching characteristics, instead of conventional large scale R-RAM story. The thermally assisted interaction between sidewall SiO_2 matrix and non-stoichiometric graded amorphous WO_x accelerates the recombination of oxygen vacancies and oxygen ions; while the thermal effect could be suppressed in monoclinic phase atomic layer deposition deposited HfO_x based memory cell since the degraded oxygen ion diffusivity. Meanwhile, the oxygen ion mobility in interstitial state of transition metal oxide matrix between filament and sidewall SiO_2 matrix also plays the dominate role of memory thermal stability. Random Telegraph Noise (RTN) testing in low resistance state of transition metal oxide R-RAM supports the statement [8-10]: random telegraph noise amplitude in low resistance state of atomic layer deposition deposited HfO_x based cell is much less than thermally oxidized non-stoichiometric WO_x based one (not shown), indicating separated oxygen ion in interstitial state of atomic layer deposition deposited HfO_x matrix is relatively stable than non-stoichiometric graded amorphous WO_x.

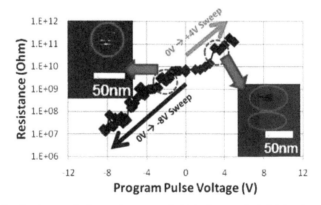

Figure 7. Conductive atomic force microscope (C-AFM) by electrical field assistance images display 5nm scaling ability of transition metal oxide based R-RAM.

SCALING LIMITATION OF TRANSITION METAL OXIDE R-RAM

Furthermore, in order to realize the scaling limitation of transition metal oxide based R-RAM, conductive atomic force microscope (C-AFM) by electrical field assistance was used to achieve both the Set and Reset programming. Both the thermally oxidized non-stoichiometric WO_x and atomic layer deposition deposited HfO_x (not shown) based transition metal oxide materials present 5nm scaling ability successfully, as shown in Figure 7. Although the images are limited by tip resolution of conductive atomic force microscope, continuous scaling down the transition metal oxide R-RAM beyond 5nm might be possible.

CONCLUSION

We have successfully demonstrated both the thermally oxidized non-stoichiometric WO_x and the atomic layer deposition deposited HfO_x based transition metal oxide R-RAM devices with sub-10nm scaling. Contrary to similar results of large R-RAM devices, scaled thermally oxidized non-stoichiometric WO_x and atomic layer deposition deposited HfO_x devices beyond 10nm technology were found to exhibit obvious differences on operation window and reliabilities. Instead of reported conventional large scale transition metal oxide R-RAM, different boundary conditions of thermally oxidized non-stoichiometric WO_x and atomic layer deposition deposited HfO_x devices plays a dominant role of R-RAM switching mechanism between 5 and 10nm.

ACKNOWLEDGMENT

This work was performed by the National Nano Device (NDL) Laboratories facilities.

REFERENCES

[1] Akifumi Kawahara, Ryotaro Azuma, Yuuichirou Ikeda, Ken Kawai, Yoshikazu Katoh, Kouhei Tanabe, Toshihiro Nakamura, Yoshihiko Sumimoto, Naoki Yamada, Nobuyuki Nakai, Shoji Sakamoto, Yukio Hayakawa, Kiyotaka Tsuji, Shinichi Yoneda, Atsushi Himeno, Ken-ichi Origasa, Kazuhiko Shimakawa, Takeshi Takagi, Takumi Mikawa, Kunitoshi Aono, "An 8Mb Multi-Layered Cross-Point ReRAM Macro with 443MB/s Write Throughput", Digest of IEEE International Solid-State Circuits Conference (ISSCC), 25-6, 2012.

[2] G. Baek, C. J. Park, H. Ju, D. J. Seong, H. S. Ahn, J. H. Kim, M. K. Yang, S. H. Song, E. M. Kim, S. O. Park, C. H. Park, C. W. Song, G. T. Jeong, S. Choi, H. K. Kang, and C. Chung, "Realization of Vertical Resistive Memory (VRRAM) using cost effective 3D Process", Digest of IEEE International Electron Device Meeting (IEDM), p.737, 2011.

[3] M. J. Kim, I. G. Baek, Y. H. Ha, S. J. Baik, J. H. Kim, D. J. Seong, S. J. Kim, Y. H. Kwon, C. R. Lim, H. K. Park, D. Gilmer, P. Kirsch, R. Jammy, Y. G. Shin, S. Choi, and C. Chung, "Low Power Operating Bipolar TMO ReRAM for Sub 10 nm Era", Digest of IEEE International Electron Device Meeting (IEDM), p.444, 2010.

[4] B. Govoreanu, G. S. Kar, Y-Y. Chen, V. Paraschiv, S. Kubicek, A. Fantini, I. P. Radu, L. Goux, S. Clima, R. Degraeve, N. Jossart, O. Richard, T. Vandeweyer, K. Seo, P. Hendrickx, G. Pourtois, H. Bender, L. Altimime, D. J. Wouters, J. A. Kittl, M. Jurczak, "10x10nm2 Hf/HfOx Crossbar Resistive RAM with Excellent Performance, Reliability and Low-Energy Operation", Digest of IEEE International Electron Device Meeting (IEDM), p.729, 2011.

[5] ChiaHua Ho, Cho-Lun Hsu, Chun-Chi Chen, Jan-Tsai Liu, Cheng-San Wu, Chien-Chao Huang, Chenming Hu, and Fu-Liang Yang, "9nm Half-Pitch Functional Resistive Memory Cell with <1 µA Programming Current Using Thermally Oxidized Sub-Stoichiometric WOx Film", Digest of IEEE International Electron Device Meeting (IEDM), p.436, 2010.

[6] ChiaHua Ho, E. K. Lai, M. D. Lee, C. L. Pan, Y. D. Yao, K. Y. Hsieh, Rich Liu, and C. Y. Lu, "A Highly Reliable Self-Aligned Graded Oxide WOx Resistance Memory: Conduction Mechanisms and Reliability", Digest of IEEE VLSI Symp., p.228, 2007.

[7] Hou-Yu Chen, Chun-Chi Chen, Fu-Kuo Hsueh, Jan-Tsai Liu, Chih-Yen Shen, Chiung-Chih Hsu, Shyi-Long Shy, Bih-Tiao Lin, Hsi-Ta Chuang, Cheng-San Wu, Chenming Hu, Chien-Chao Huang, and Fu-Liang Yang, "16nm Functional 0.039 ☐m2 6T-SRAM Cell with Nano Injection Lithography, Nanowire Channel, and Full TiN Gate" Digest of IEEE International Electron Device Meeting (IEDM), 28.7, p.958, 2009.

[8] J.-K. Lee, H. Y. Jeong, I.-T. Cho, J. Y. Lee, S.-Y. Choi, H.-I. Kwon, and J.-H. Lee, IEEE Electron. Device. Lett. 31, 603 (2010).

[9] ChiaHua Ho and Fu-Liang Yang, "Overview of Metal-Oxide Resistive Memory" in "Nonvolatile Memories: Materials, Devices and Applications'" American Scientific Publishers, ed T.Y Tseng and S.M Sze (2012).

[10] G. Bersuker, D. C. Gilmer, D. Veksler, J. Yum, H. Park, S. Lian, L. Vandelli, A. Padovani, L. Larcher1, K. McKenna, A. Shluger, V. Iglesias, M. Porti, M. Nafría, W. Taylor, P. D. Kirsch, and R. Jammy, "Metal Oxide RRAM Switching Mechanism Based on Conductive Filament Microscopic Properties", Digest of IEEE International Electron Device Meeting (IEDM), p.456, 2010.

[11] N. F. Mott, et al., Electronic Processes in Non-Crystalline Materials, Oxford: Clarendon Press, 1979, pp 28-33, p.37.

Author Index